Turn On and Tune In

Psychedelics, Narcotics and Euphoriants

Turn On and Tune In
Psychedelics, Narcotics and Euphoriants

John Mann
Emeritus Professor of Chemistry, Queen's University Belfast, UK

RSCPublishing

ISBN: 978-1-84755-909-8

A catalogue record for this book is available from the British Library

Published by The Royal Society of Chemistry,
Thomas Graham House, Science Park, Milton Road,
Cambridge CB4 0WF, UK

Registered Charity Number 207890

For further information see our web site at www.rsc.org

Preface

Timothy Leary's exhortation to *'turn on, tune in, drop out'* was part of his campaign to encourage the use of drugs like LSD and psilocybin for the expansion of the mind; and many people in the 1960s needed little encouragement to use these and other mind-expanding drugs. But in reality this was nothing new since humans have been experimenting with mind-altering chemicals for several millennia and this is appositely described by Louis Lewin, the German pharmacologist, in his 1924 book entitled *Phantastica: Narcotic and Stimulating Drugs*. He wrote:

From the first beginning of our knowledge of man, we find him consuming substances of no nutritive value, but taken for the sole purpose of producing for a certain time, a feeling of contentment, ease and comfort.

He classified agents capable of changing cerebral function into five main classes:

- Euphorica – sedatives of mental activity which produce a state of physical and mental comfort, *e.g.* opium and cocaine.
- Phantastica – hallucinating substances, *e.g.* mescaline, cannabis and scopolamine.
- Inebriantia – causing cerebral excitation followed by depression, *e.g.* alcohol, chloroform, ether and benzene.

Turn On and Tune In: Psychedelics, Narcotics and Euphoriants
By John Mann
© John Mann 2009
Published by the Royal Society of Chemistry, www.rsc.org

- Excitantia – mental stimulants, *e.g.* caffeine and nicotine.
- Hypnotica – soporifics, *e.g.* chloral.

Lewin could not write about LSD since it had not been discovered, and the large-scale recreational use of many of these drugs was to be a twentieth-century phenomenon which he could not anticipate.

There is a continuing fascination with psychedelic, narcotic and euphoriant substances, and this book attempts to provide a history of their discovery and use, with an emphasis on the social and anthropological aspects of this use and also abuse. In particular, the many colourful characters who have been involved in drug production or use are given prominence. Since all of these substances are chemicals with biological activities, the basic chemistry and pharmacology is also described. In fact this is a key feature of the book since, although there are a number of excellent books dealing with individual drugs (cocaine, cannabis, opium, *etc.*), none of these covers all aspects of the drugs from their historical and social importance through to their chemistry and pharmacology. Nonetheless the book is intended to be an enjoyable read rather than a textbook – which it very definitely is not. So although there is a glossary at the back with key chemical structures and other core scientific information, the general reader can ignore this section without any loss.

Jerry Garcia of the Grateful Dead is alleged to have said that if you could remember the 1960s you weren't actually there – though this comment has also been attributed to several other pop stars of the era. But this is to miss the point about the use of mind-altering drugs, since their effects on the brain are but one of the many areas of interest ascribed to these multifaceted molecules. Their impact on history, politics and social structure are both significant and fascinating, and although most of the substances are tightly controlled, many of them are still widely used. One only has to consider the fact that the war in Afghanistan is probably unwinnable all the time opium is the main crop of that country to appreciate the impact of mind-altering drugs on world affairs.

As usual there are people to thank and I received enthusiastic support and advice for this project from Professor Phil Page, the staff at RSC Publishing, and my wife Rosemary, who has tried very hard to steer me away from scientific excess so that the general reader can enjoy the story.

Contents

Turn On and Tune In: Psychedelics, Narcotics and Euphoriants
By John Mann
© John Mann 2009
Published by the Royal Society of Chemistry, www.rsc.org

From Ergotism to LSD

Timothy Leary's advice to '*turn on, tune in, drop out*' was a 1960s exhortation to experiment with LSD and other psychedelics, but humans had been consuming ergot alkaloids related to lysergic acid diethylamide for at least a thousand years, and perhaps since the time of the ancient Mesopotamian empires. The fertile alluvial lands that lie between the mighty Euphrates and Tigris rivers were the cradle of some of the great early civilisations including those of the Sumerians, Assyrians and Babylonians. These large and sophisticated nations developed agriculture on a vast scale with huge annual harvests of wheat and barley, and perhaps rye (which was probably first domesticated in what is now Turkey at least 4000 years ago). For the first time in the Middle East, a well-developed leadership and infrastructure could harness the energies of a large peasant workforce for irrigation projects that provided large annual grain surpluses. These could support a non-nomadic lifestyle and with it the spare time to develop new crafts and written languages. And it is on an ancient Assyrian cuneiform tablet from around 600 BC that we first learn of the '*noxious pustule in the ear of grain*'. This may refer to the purplish black sclerotium (the over-wintering form) of the ergot fungus *Claviceps purpurea*, which resembles a cock's spur (*ergot* in French) (see Figure 1.1). Crop damage is also mentioned in the bible:

'*I smote you with blasting and with mildew* . . . ' (Haggai, **2**, 17); and

Turn On and Tune In: Psychedelics, Narcotics and Euphoriants
By John Mann
© John Mann 2009
Published by the Royal Society of Chemistry, www.rsc.org

Figure 1.1   The ergot fungus growing on rye – note the spur-shaped dark sclerotium.
© Blackthorn Arable.

'If there be in the land famine, if there be pestilence, blasting and mildew . . . ' (Kings, **8**, 37).

However, although the serious health problems associated with consumption of contaminated grain were undoubtedly known in ancient times, we do not know if the link to the ergot fungus was made. We do know that rye – the usual grain host for the fungus – was not introduced from Turkey into Europe before about 1800–1500 BC, and we have to wait for the reports of the mediaeval chroniclers for our first glimpse of ergotism. Perhaps the earliest description can be found in the diaries (the *Annales*) for AD 857 from the convent of Xanten close to Duisberg on the Lower Rhine in Germany. These relate that *a great plague of swollen blisters consumed the people by a loathsome rot so that their limbs were loosened and fell off before death*. The gangrene that is so graphically depicted here was one of the two types of ergotism – the gangrenous type – while the other form was most often described as convulsive ergotism and led to so-called 'dancing epidemics'. The chroniclers agree on the range of symptoms associated with these two disease types. Gangrenous ergotism began with itching and a crawling sensation in the feet with associated sensations of hot and cold in the extremities leading to extensive blistering and gangrene of hands and feet. Loss of these limbs was sometimes followed by recovery but death from septicaemia was the more common sequel. The convulsive form of ergotism was associated with effects on the central nervous system giving rise to tingling in the hands and feet, convulsive movements of the muscles giving rise to staggering and involuntary movements – hence the 'dancing epidemics'.

Many of the chronicles describe plagues giving rise to symptoms that might well be ascribed to ergotism. For example, the account of Frodoard of an *ignis plaga* (fire plague) in Paris around AD 945 which attacked various limbs of the body (*diversa membra ignis plaga per-adunta*) with relief only coming with death of the sufferer. Though Frodoard goes on to report that some were saved by the local leader Hugo who supplied daily rations of (presumably) bread free of ergot contamination:

Asserantur ab hoc peste salvatii (saved from the plague): quos Hugo quoque Dux stipendiis aluit quotidianis (daily rations from Duke Hugo).

Salvation was also provided by the local clergy and there are various accounts of plagues in Aquitaine and Limousin during the period AD 950 to 1000 where the bones of St Martial were shown to the victims with positive results. One can only assume that these clergy had supplies of uncontaminated bread to give to the victims, thus reversing the effects of the ergotism in those lightly affected. Alternative saintly relics were used in other regions including those of St Génévieve in Paris and St Martin in Tours; but it was the remains of St Anthony that were most widely touted as an effective means for the cure of holy fire or *ignis sacer*.

The origins of the myth of St Anthony, like much of the early history of ergotism, have to be viewed with a degree of scepticism. It is usually accepted that he was born near Koma in Egypt around AD 250 into a wealthy Christian family. He was converted to the life of a hermit fol-lowing the death of his parents, and by AD 270 had given away all of his worldly goods and had retreated to a tomb near his village. Many paintings show him tormented by visions of wild beasts during his period of seclusion, though he also seems to have been tormented by people seeking his advice (see Figure 1.2), and he eventually retreated to the depths of the Sinai desert and lived in an old abandoned fort. After about 15 years of contemplation and prayer, he emerged to found a cult of Christian monasticism, and spent the rest of his long life devoted to these religious activities. Following his death, probably in AD 356, his remains were hidden locally only to be discovered some years later and then transferred to St John's Church in Alexandria. Here

Figure 1.2   An imagined meeting of St Anthony and a patient suffering from ergotism
– note the amputation and the burning hand. © Wellcome Images.

they lay for several centuries but they were transferred to Constantinople before the Arabs overwhelmed Alexandria in AD 641. They
remained undisturbed in the Church of St Sophia for about
400 years.

   After this the story becomes even more hazy though it is usually
assumed that what little remained of him was taken to France by crusaders returning from Constantinople. Jocelin, the Count of Dauphiné,

is usually credited with taking the relics back to the Dauphiné region in south-eastern France – but the truth remains elusive. The location of the church where the remains were deposited around AD 1070 is also in doubt, though the one in the tiny village of St Didier de la Motte (now St Antoine l'Abbaye, near Marcellin, Isère) is most often cited in the chronicles. What seems more certain is that the first hospital dedicated to the treatment of victims of ergotism was erected nearby in around AD 1090 by Gaston, a local nobleman whose son Guerin had been cured of ergotism following a pilgrimage to St Didier de la Motte. This hospital with its team of dedicated brothers of the order of St Anthony served as a model for dozens of other hospitals that sprang up all over France (*e.g.* in Besançon and Chambéry). The great significance of the association of St Anthony with ergotism has been celebrated in many fine paintings and other religious artefacts, most notably the wonderful sixteenth-century altar painting by Matthias Grünewald in the church at Colmar, a small town between Strasbourg and Basel.

Obviously all of these activities occurred in the depths of the Middle Ages, and it is worth looking at some of the more reliable medical and historical records to try to untangle fact from fiction. One fact quickly emerges and this concerns the greater prevalence of the necrotic form of ergotism west of the Rhine, primarily in France, while the convulsive type – often mistaken for epilepsy – seems to have been more common in Germany and Russia. Presumably these differences correspond to the levels of the various ergot alkaloids present in the geographically distinct strains of fungus, and to the mode of preparation or cooking of the bread. The major ergot alkaloid is usually ergometrine, which is a potent vasoconstricting agent and this would be expected to have a significant effect on the blood vessels of the extremities (hence the gangrene) and on neurotransmission within the central nervous system (hence the neurological effects). The vasoconstrictive effect was certainly recognised and used as early as the sixteenth century when the German physician Adam Lonitzer described the use of the sclerotia of ergot to stimulate contraction of the uterus and thus 'quicken labour'. In his Kräuterbuch (four editions between 1557 and 1577) – actually the title was *Book of Herbs and Artificial Counterfeiting, With the Art of Distillation* – he provided specific instructions for this new medication: three sclerotia to be administered with repeat doses if necessary.

A more scientific description was given by the German physician Paulizky in 1787 in a paper to the *Neues Magazin für Arzte*, and here he described his extract of ergot as *pulvis ad partum* as providing a more rapid and powerful quickening of labour than any other known drug. Crude ergot became very popular amongst the midwives and physicians in much of Europe, though it was less enthusiastically adopted in America. So, although John Stearns, a physician from Saratoga County in New York State, extolled the virtues of administration of 5–10 g of crude ergot to produce a rapid delivery of the baby, he did warn of the severe adverse effects of nausea and vomiting and admitted that the uncertain constitution of the extracts could affect the outcome. Writing to a Mr Akerly in January 1807 he says:

> *In compliance with your request I herewith transmit you a sample of pulvis parturiens, which I have been in the habit of using for several years, with the most complete success. My method of administering it is either in decoction or powder. Boil half a drachm of the powder in half a pint of water, and give one third every twenty minutes till the pains commence . . . If the dose is large it will produce nausea and vomiting. In most cases you will be surprised with the suddenness of its operation; it is, therefore, necessary to be completely ready before you give the medicine, as the urgency of the pains will allow a short time afterwards.*

And he went on to discuss the source of his material:

> *It is a vegetable, and appears to be a spurious growth of rye. On examining a granary where rye is stored, you will be able to procure a sufficient quantity from among the grain. Rye which grows in low, wet ground, yields it in greatest abundance.*

Oliver Prescott produced a pamphlet in 1813 for the Massachusetts Medical Society along the same lines, and such was the enthusiasm for ergot that it was included in the US Pharmacopoeia in 1820. However, other physicians were more cautious and as early as 1824 the New York physician Hosak had warned of the serious risk of rupture of the uterus and death of the mother. He even suggested that crude ergot should be renamed *pulvis mortem*, and the rising toll of deaths resulting from its use provided the death knell for ergot, at least in America, and by the end of the nineteenth century it was no longer in use.

Two strange occurrences often blamed upon the neurological effects of ergotism are worth considering, even if the links to contaminated rye have never been conclusively established. The first concerns *la Grande Peur*, or the Great Fear, which gripped whole cohorts of the French peasantry for about 18 days at the end of July 1789. There is little discernible pattern to the riots in which the peasants rose up against the tyranny of the French landowners who controlled almost every facet of their lives. It is suggested that they were responding to rumours that brigands from the north had been sent to steal or damage their crops. Certainly the concurrent agitation against the aristocracy that was occurring in Paris and its environs, which marked the start of the French Revolution (the Bastille had fallen on July 14th), had given rise to widespread fear and uncertainty amongst the peasant class. In regions as far apart as Normandy, Dauphiné, Provence and Aquitaine, the peasants looted and burned local châteaux but (for the most part) stopped short of murder, and most eyewitness accounts imply that this was uncoordinated madness rather than a revolution. A tentative link between the various outbreaks of vandalism and ingestion of ergot-contaminated rye has been made. In particular, the weather records for 1788–1789 reveal that a cold winter was followed by a cold and damp spring, which would have been ideal growing conditions for the fungus. Early summer was then warm and dry, favouring airborne spread of the fungal spores, and the warm and wet summer completed a weather cycle that was hugely favourable for growth of the sclerotia. To make matters worse, the previous year's harvest had been poor so the peasants were probably less assiduous in their cleaning of the rye harvest. Of course this regional madness brought on by ergot-contaminated rye could be pure fantasy apart from the fact that certain regions of France escaped this epidemic. One of these was the Sologne which had been badly affected by the gangrenous form of ergotism throughout the Middle Ages, so the peasants could be expected to be more careful in rooting out ergot from the harvested rye. We will never know what caused *la Grande Peur* but it is clear that the landowning aristocracy failed to respond to the genuine grievances that were voiced, and they soon faced a much larger conflagration which had nothing to do with ergotism – the French Revolution.

The second episode of unexplained behaviour concerns the Salem witchcraft trials of 1692. Starting in December 1691 a number of villagers began to behave in a peculiar way suffering from what were described as 'distempers with disorderly speech' and 'odd postures and convulsive fits', and these led to a suspicion of witchcraft. At their trial some of the defendants complained of hallucinations and 'crawling sensations in the skin', and all of these symptoms could, in principle, be ascribed to ergotism. Rye was certainly the most successful cereal grown by the New England settlers, and as with *La Grande Peur*, the weather records for 1691 reveal early spring rains followed by a hot and stormy summer which could have encouraged the growth of *Claviceps purpurea*. Harvesting and threshing of the crop would have been over by Thanksgiving Day, so the bread baked in December would have been the first to contain ergot if indeed it was present. Whatever the cause of the strange behaviour – and the evidence for ergotism is highly tenuous – by September 1692, 20 villagers had been found guilty of witchcraft and executed.

As with many other biologically interesting natural extracts, pharmacists and chemists began to investigate the chemical constituents of ergot early in the nineteenth century. Progress was slow and the first supposedly pure constituent – ergotinine – was not isolated until 1875 by the French pharmacist Charles Tanret, though this proved to be a mixture of alkaloids with little discernible biological activity. It took another 30 years before George Barger and Francis Carr of the Wellcome Research Laboratories in London managed in 1906 to isolate what they believed to be the pure and biologically interesting ergot alkaloid they christened ergotoxine, though this was subsequently shown to be a mixture of active alkaloids, mainly ergocornine. This was followed 12 years later by Arthur Stoll's isolation of ergotamine (in 1918) at Sandoz in Basel, though it took until 1951 before he and Albert Hofmann established the correct chemical structure. The trio of major alkaloids was completed by perhaps the most medicinally interesting of the ergot alkaloids – ergometrine – isolated by Harold Dudley and John Moir of the National Institute for Medical Research in London in 1935, and independently by Morris Kharasch in Chicago, Marvin Thompson in Maryland and Stoll and Burkhardt at Sandoz. These researchers could not agree on one name, so ergometrine in Europe was called ergonovine in the USA. This was shown to have a pronounced effect on uterine smooth muscle producing increased muscular tone with practical value in the prevention and treatment of postpartum haemorrhage (PPH). Since postpartum haemorrhage is still a major factor responsible for the deaths of as many as 150,000 women each year, especially in developing countries, the introduction of ergometrine for treatment and prevention of PPH in 1935 was a major lifesaver. However, over the last 70 years of use, obstetricians have gradually moved away from ergometrine towards the peptide hormone oxytocin since this also stems PPH but with a more acceptable profile of side-effects for the patient.

Another clinically important alkaloid is ergotamine which was introduced in the mid 1920s by Ernst Rothlin, a colleague of Arthur Stoll, for the therapy of moderate to severe attacks of migraine. Rothlin took the bold decision to inject the alkaloid subcutaneously into two patients with severe migraine who had not responded to any treatment, and fortunately achieved immediate success. A proper clinical evaluation was then undertaken by H. W. Maier at the Burgholzli Hospital in Zurich and he was able to confirm the efficacy of ergotamine.

The alkaloid appears to act by vasoconstriction (narrowing) of blood vessels associated with the carotid artery where vasodilation has been produced by over-production of the neurotransmitter serotonin (5-hydroxtryptamine or 5-HT). However, unlike ergometrine which exerts its effects mainly at 5-HT receptors, ergotamine is a non-selective drug and interacts with various neurotransmitter receptors including those for noradrenaline and dopamine. Not surprisingly, prolonged administration of the drug (the maximum recommended dose is 6 mg/day) can give rise to the classic symptoms of ergotism including severe peripheral vasoconstriction leading to gangrene.

All of this interest in the clinical utility of ergot alkaloids necessitated exploitation of various sources of the compounds. Sandoz introduced ergotamine tartrate in 1921, where the ergotamine was isolated directly from the fungus. More recently, a number of other pharmaceutical companies including Novartis (which took over Sandoz), Boehringer Ingelheim, Eli Lilly and Farmitalia all produce ergot alkaloids either by fermentation technology (around 60% of the supply) or from cultivation of triticale, which is a hybrid of wheat and rye. All of these natural products are complex amide derivatives of lysergic acid, and most of this supply of natural ergot alkaloids is converted into lysergic acid through hydrolysis. This parent acid had already been prepared in 1934 by Jacobs and Craig at the Rockefeller Institute in New York through hydrolysis of ergotinine with methanolic potassium hydroxide solution.

The first total synthesis of lysergic acid was accomplished by R. B. Woodward and his coworkers in 1954, and this was characterised by both the elegance of the approach and its use of simple reagents. Ergotamine was first synthesized in 1961 by Albert Hofmann – discoverer of LSD – and all of the other ergot alkaloids can be produced from the acid chloride of lysergic acid (or other activated derivative) and the corresponding amines. Of the semi-synthetic compounds that have been introduced into clinical use, methysergide, which is a methylated derivative of ergometrine, has probably been the most successful and is used for the treatment of patients with recurrent vascular headaches.

The only other ergot alkaloid with any significant clinical utility is bromocriptine and this is synthesised from lysergic acid. It possesses dopamine-like activity and has useful activity when given by mouth to patients with Parkinson's disease, which is due in part to a deficiency in dopamine. The drug is not as effective as L-DOPA (3,4-dihydroxyphenylalanine) – the standard drug for the condition – and bromocriptine is usually reserved for patients who no longer respond to

L-DOPA. Another less complex analogue pergolide was introduced by Eli Lilly in 1979, and when given in conjunction with L-DOPA the drug combination not only provides a great lessening of the classic symptoms of Parkinson's but also extends the period of good quality life before further neuorodegeneration occurs.

But by far and away the most famous ergot alkaloid is, of course, the diethylamide of lysergic acid – LSD. In 1938, Stoll and Hofmann were working on a series of simple amides of lysergic acid, one of which was the diethylamide. This had some structural similarity to the diethylamide of nicotinic acid which was a successful analeptic drug (a circulatory and respiratory stimulant) called nikethamide or coramine. Initial pharmacological analysis of LSD showed that it had strong oxytocic activity causing strong uterine contractions with about 70% of the activity of ergometrine. Nothing further was done with the compound until 1943 when Hofmann was asked to prepare further quantities of LSD for additional pharmacological studies. On Friday 16th April he was recrystallising a small quantity of the alkaloid when he was overcome by a sense of restlessness and vertigo. In his book *LSD: My Problem Child*, he recalled these effects as written in the report that he submitted to Stoll after the weekend:

> *I was forced to interrupt my work in the laboratory in the middle of the afternoon and proceed home, being affected by a remarkable restlessness, combined with a slight dizziness. At home I lay down and sank into a not unpleasant intoxicated-like condition, characterized by an extremely stimulated imagination. In a dreamlike state, with eyes closed, I perceived an interrupted stream of fantastic pictures, extraordinary shapes with intense, kaleidoscopic play of colours. After some two hours this condition faded away.*

Thinking, not unreasonably, that these effects had been caused by accidental ingestion of LSD, he decided to try a measured dose of the compound. On Monday 19th April he deliberately swallowed 0.25 milligram of LSD which was many times higher than the minimum dose required to produce marked behavioural changes. Not surprisingly the subsequent hallucinogenic experience was very dramatic and he had to ask his laboratory assistant to help him get home. Due to wartime restrictions on the use of cars, they had to travel by bicycle and by all accounts the journey was terrifying, but at least they reached his home in one piece, which might not have been the case if he had driven. Once home, he collapsed on the sofa with an urgent request to summon the family doctor and fetch milk to use as an antidote. In his book he tells of his experience:

> *Everything in the room spun around, and the familiar objects and pieces of furniture assumed grotesque, threatening forms. They were in continuous motion, animated, as if driven by an inner restlessness. The lady next door, whom I scarcely recognised, was no longer Mrs R, but rather a malevolent, insidious witch with a coloured mask.*

The doctor arrived and after examination reassured him that all his vital signs were normal though his pupils were extremely dilated. After drinking copious quantities of milk, Hofmann began to relax a little and no longer feared for his life. He even began to enjoy the experience:

> *Kaleidoscopic, fantastic images surged in on me, alternating, variegated, opening and then closing themselves in circles and spirals, exploding in coloured fountains . . . Every sound generated a vividly changing image, with its own consistent form and colour.*

After several hours he fell into an exhausted sleep but awoke next morning without any kind of hangover and in fact experienced *a sensation of well-being and renewed life flowed through me.* Ernst Rothlin, now director of the Pharmacology Department at Sandoz, was initially sceptical about Hofmann's report, but two of Hofmann's colleagues repeated the experiment but with smaller doses and reported their own fantastic experiences. Sandoz could hardly exploit the hallucinogenic properties of LSD, and the company tried very hard to find a therapeutic use for the drug. They made LSD freely available to qualified investigators and for a while it looked as if the drug might have utility in the treatment of some psychiatric conditions like schizophrenia and also as a treatment for alcoholism. Unfortunately, careful experimentation in the 1950s was often side-lined by sensational magazine articles like the one by Sydney Katz entitled *My 12 hours as a madman* after he participated in a Canadian LSD study. In addition, over the intervening years, there have been as many reports of adverse reactions to the drug as positive responses, with real dangers for drug takers who experience feelings of omnipotence or invulnerability, and many suicides have resulted from so-called LSD psychosis. The apparent discovery that chromosomal abnormalities were commonly found in patients after several months of treatment also produced a very negative view of the drug. However, these clinical failures did nothing to dampen the enthusiasm of the emerging hippy culture of the 1960s which eagerly embraced the drug.

Contemporaneous with these clinical experiments, Hofmann made further surprising discoveries. He was firstly instrumental in bringing to light the South American use of seeds of the plant *Rivea corymbosa* which had great significance in Aztec culture. The story starts in the late 1930s when the renowned ethnobotanist Richard Evans Schultes – later to become Director of the Botanical Museum in Harvard University – first suggested that the Aztec magical preparation *teonanacatl* had been an extract of various species of *Psilocybe* mushrooms. This magical preparation had been described by Bernardino de Sahagun, a Franciscan friar who accompanied Hernan Cortes on his bloody campaign of conquest of South America in the sixteenth century. In his report of what he encountered, entitled *Historia General de las Cosas de Nueva España*, he wrote about a mushroom party:

> *And they ate the mushrooms with honey – when the mushrooms were taking effect, there was dancing, there was weeping . . . some saw in a vision that they would die in war, some saw in a vision that they would be devoured by wild beasts . . . some saw in a vision that they would become rich and wealthy.*

Other contemporary writers also refer to mushroom ceremonies and numerous stone sculptures in the form of mushrooms have been unearthed, many bearing carvings that depict gods or demons, and some of these can be dated as far back as 500 BC (see Figure 1.3).

Schultes' assertions were complemented by those of another famous ethnobotanist, R. Gordon Wasson, who together with his wife Valentina made many further discoveries concerning the magic mushroom cult associated with *teonanacatl*. In particular, during travels in Mexico in 1955, they managed to track down a village shaman (*curandera*), Eva Mendez, who was prepared to allow them to sample *teonanacatl*. In a long feature in *Life* magazine in June 1957 he told of his experiences:

> *The visions came . . . they emerged from the center of the field of vision, opening up as they came, now rushing, now slowly, at a pace that our will chose. They were in vivid colour, always harmonious. They began with art motifs, angular such as might decorate carpets or textiles . . . then they evolved into palaces with courts, arcades, gardens . . . resplendent palaces all laid with semiprecious stones. Then I saw a mythological beast driving a regal chariot.*

Figure 1.3   Albert Hofmann with a pre-Columbian 'mushroom stone'. © Barnabas
Bosshart/Corbis.

Wasson tried the mushrooms on several occasions and always experienced similar visions, and it was quite clear from his many discussions with the shaman (she was a *curandera de primera categoria*, of the highest quality) that these ceremonies were a kind of 'holy communion' with the spirits. They were not part of a healing process for sick people but rather a route by which the participants could seek answers to questions through their visions. Wasson perceived this as a fascinating example of pre-Christian practices merging with the new ways of the Catholic Church that had been introduced by the conquistadores and the Franciscan and Benedictine missionaries who accompanied them. Wasson also noted that Eva Mendez had been consuming the mushrooms on a more or less regular basis for 35 years with no apparent ill effects.

On a subsequent trip, the Wassons were accompanied by a French mycologist, Robert Heim, and he subsequently managed to cultivate the most important mushroom species, *Psilocybe mexicana*, in his laboratories at the Musée Nationale d'Histoire Naturelle in Paris and sent samples on to Hofmann at the Sandoz laboratories in Basel. Unfazed by his earlier experiences with LSD, Hofmann consumed 32 dried mushrooms (about 2.4 grams), which was reckoned to be a typical dose for the South American Indians, and then recorded his experiences in his laboratory notebook:

Thirty minutes after taking the mushrooms, the exterior world began to undergo a strange transformation. Everything assumed a Mexican character . . . Whether my eyes were closed or open I saw only Mexican motifs and colours. When the doctor supervising the experiment bent over to check my blood pressure, he was transformed into an Aztec priest and I would not have been astonished if he had drawn an obsidian knife . . . At the peak of the intoxication, the rush of interior pictures, mostly abstract motifs rapidly changing in shape and colour, reached such an alarming degree that I feared that I would be torn into this whirlpool of form and colour and would dissolve. After about six hours the dream came to an end.

Hofmann went on to isolate and characterise the main psychoactive constituents of *Psilocybe mexicana* and these proved to be mainly psilocybin with a little psilocin, both of which have close structural similarity to the neurotransmitter 5-HT and block the effects of 5-HT. Psilocybin has only about 1% of the potency of LSD and use of *Psilocybe* mushrooms by the hippy culture never achieved the star status of LSD, though these magic mushrooms are still very much part of the way of life of, for example, the Mazatec Indians in Mexico.

A further surprise was in store for Hofmann in 1959 when R. Gordon Wasson sent him seeds of *Rivea corymbosa*, a species of Morning Glory, that had been collected in the state of Oaxaca in Mexico (see Figure 1.4). Schultes had already written about this plant in 1941 in his publication for the Botanical Museum of Harvard entitled *A Contribution to Our Knowledge of Rivea Corymbosa: the Narcotic Ololiuqui of the Aztecs*. A much earlier description of ololiuqui had been written by Francisco Hernandez, physician to Philip II of Spain, during his travels in South America during the period 1570–1575. In his *Rerum Medicarum Novae Hispaniae Thesaurus* he described the preparation of ololiuqui and noted:

When the priests wanted to commune with their Gods and receive a message from them . . . they ate this plant and a thousand visions and satanic hallucinations appeared to them.

The term ololiuqui or ololiúhqui was a Nahuatl word meaning 'round thing' which referred to the seeds.

To Hofmann's amazement, the major chemical constituents of *Rivea corymbosa* seeds were derivatives of lysergic acid with an uncanny structural resemblance to LSD, though these compounds subsequently proved to be about 20 times less potent and were mostly narcotic rather than hallucinogenic. When Hofmann first presented his results at the IUPAC Congress on Natural Products in Sydney in 1960, there was

Figure 1.4   *Rivea corymbosa* – source of the Aztec magical preparation. © USDA.

considerable scepticism that similar compounds could be found in a plant and a fungus, since these types of organism are very well separated in evolutionary history. Nonetheless Hofmann had now been responsible for establishing the involvement of ergot alkaloids in the cultures of peoples on two continents separated by the vast expanse of the Atlantic Ocean.

He and Wasson had one final piece of ethnopharmacological detective work to accomplish and that was their attempt to explain the Eleusinian magical ceremonies of Ancient Greece. For this they elicited the help of the Greek scholar Carl Ruck, who was an expert on the 4000-year-old Eleusinian mysteries which had taken place each autumn outside Athens. These were purported to involve a dramatic reconstruction of the myth of Persephone, who was drugged then abducted to the underworld, but subsequently returned in triumph with a son she had conceived in Hades. Details of the rituals involved were never documented because the participants were sworn to secrecy on pain of death, but Ruck had assembled enough fragmentary information to be persuaded that the participants were drugged with a concoction of wine laced with an extract of barley. Rye was not cultivated in ancient Greece but both wheat and barley were common crops and *Claviceps purpurea* can grow on both of these cereals. It took little more to convince Hofmann and Wasson that the Eleusinian mysteries were nothing less than magical rituals akin to those practised by the Aztecs using ololiuqui, though the evidence is very thin indeed. What gives it some credence is the fact that ergometrine is the only ergot alkaloid that is easily extracted from the sclerotia using water (or presumably wine), and as usual Hofmann experimented on himself by consuming about 2 mg of ergometrine and did experience hallucinations. Whatever happened in Eleusius this again provided a graphic demonstration of consumption of ergot-contaminated grain leading to the convulsive form of ergotism so often observed in the Middle Ages.

These primarily scientific activities caught the imagination of a growing number of psychiatrists who speculated that consumption of magic mushrooms or LSD might help patients to reveal their innermost problems. One of the prime movers in this area was a young Harvard academic called Timothy Leary who did more than anyone else to popularise the cult of LSD. Within a few years his enthusiasm for this drug would earn him the opprobrium of much of the US hierarchy including President Nixon, who (prior to his own impeachment!) labelled him the *most dangerous man in America*. Leary had completed a Ph.D. in psychology at the University of California at Berkeley in 1950, and became, in 1955, the Director of Psychiatric Research at the Kaiser Family Foundation Hospital. He even wrote a 518-page textbook entitled *The Interpersonal Diagnosis of Personality*, which received very favourable reviews. At the end of 1959 he accepted a position as Assistant Professor at the Harvard Center for Personality Research and appeared to be on track for a promising academic career. Until, that is, he spent a vacation with colleagues Richard Alpert, Anthony Russo and other mutual friends in a small Mexican village called Cuernavaca. Evidently the conversation turned to thoughts of magic mushrooms and they managed to locate a local Indian *curandera*, who sold them a bag of mushrooms which apparently contained psilocybin. The resultant experience changed Leary's life and he returned to Harvard determined to carry out scientific research in a new research group that he christened the Harvard Psilocybin Research Program – he even persuaded Harvard to fund a project that sought to examine the clinical effects of psilocybin. Sandoz very kindly agreed to supply the drug, which had been isolated as a result of Hofmann's earlier studies, and Leary and his collaborators now had to plan their experiments carefully since this was uncharted territory. Well almost, because at around this time Aldous Huxley – author of the highly influential *Brave New World* (1932) – came to Massachusetts Institute of Technology as a visiting professor. Back in 1953 he had experimented with mescaline from the peyote cactus – yet another of the Mexican magical preparations commonly know as *peyotl*, and one that had been investigated in the clinical setting of the Maudsley Hospital in London during the 1930s. Huxley had written about his experiences with mescaline in his 1954 book *The Doors of Perception* and it is quite clear from the text that he considered mescaline to be a force for good since it allowed the user to gain insight into the world around them.

While at MIT, he and Leary spent some time together discussing Leary's proposed research programme. Both apparently had firm views that the correct use of psychedelics could divert humans from their warlike and other violent preoccupations, and this led Leary to carry out psilocybin experiments, not only with students and colleagues at Harvard, but also within prison populations, to see if psychedelics could reduce the level of re-offending once the prisoners were released. Various pseudo-religious investigations were also carried out at this time to compare the psychedelic experience with those described in the bible and elsewhere where people reported having visions and claimed to have experienced 'ecstasy'. Not surprisingly the Church began to take notice of this academic who claimed that a pill could produce experiences hitherto claimed to be achievable only when one came close to God. But Leary was about to ruffle more feathers as he embarked upon a programme of research that would cause fear and consternation not only in the Church but all the way to the White House.

Leary is reputed to have been introduced to LSD by an Englishman, Michael Hollingshead, an equally colourful character who, it is claimed, obtained a gram of LSD from Sandoz simply by writing to the company using official notepaper from a New York Hospital and claiming that he needed the drug for medical experiments. Whatever the truth of this claim, Hollingshead visited Leary at Harvard and persuaded him to take his first trip with LSD.

Leary quickly became something of an evangelist for LSD and criss-crossed America giving lectures and talking to influential people trying to impress on them the ways in which LSD allowed people to explore the realities of their lives by 'reprogramming their brains'. His extended absences from Harvard, not to mention the growing pressure from the parents of students who were alarmed by bizarre stories of LSD use by their offspring, eventually led to Leary being sacked from his position in 1963. He seems to have been unfazed by this, and was anyway too busy setting up his International Foundation for Internal Freedom in Cambridge, Massachusetts, and a kind of summer school in Mexico, for those who wanted to experience the mind-expanding possibilities offered by LSD. However, the centre in Mexico only survived for one summer before being closed by the Mexican police, and Leary and his team were deported, and to make matters more difficult for Leary, the Federal

Drug Administration now decided that LSD was too dangerous to be used in a non-clinical setting, so obtaining the drug from Sandoz became highly controlled and the company stopped manufacturing both LSD and psilocybin in August 1965. The CIA had also now taken a keen interest in Leary's activities, not least because they had been carrying out their own investigations on the utility of LSD as an incapacitating agent for the battlefield. The actual extent of these activities only came to light during a Senate Enquiry in 1977 which revealed a plethora of covert activities on chemical and biological weapons covered by the codename MK-Ultra. The revelations about experiments on prisoners and mentally ill patients using LSD were, of course, defended on the grounds of national security during the period of the Cold War with Russia and her allies.

Having lost his job and his summer school, Leary had to find a new base for his activities and once again his luck held, and he was allowed to set up an experimental commune in a 64-room mansion at Millbrook in New York State. This gothic pile was used by the grandchildren of the immensely rich Mellon family, who were persuaded of the benefits of LSD, and it was here, and later on a trip to India, that Leary formulated his new religion, which he christened the League of Spiritual Discovery. There were just two commandments:

> *Thou shalt not alter the consciousness of thy fellow man.*
> *Thou shalt not prevent thy fellow man from altering his own consciousness.*

And the manifesto – *we solemnly publish and declare that we are free and independent, and that we are absolved from all allegiance to the United States Government and all governments controlled by the menopausal* – did little to endear Leary to the Government, which viewed him as a subversive. A view that was confirmed when he gave his now-famous speech to the huge crowd of around 30,000 attending the 'Human Be-In' held in San Francisco in January 1967. The speech contained his immortal appeal:

> *The only way out is in – turn on, tune in, drop out! Of High School, junior executive, senior executive. And follow me! The hard way!*

Subsequently Leary sought to explain this apparently negative advice. He claimed that 'turn on' implied an activation of the mind through drug use; 'tune in' meant achieving harmonious interaction with the world and 'drop out' meant detachment from the world. So the

Figure 1.5   Timothy Leary at one of his rallies enthusing about the benefits of LSD and other drugs. © Bettmann/Corbis.

exhortation was not at all subversive. Whatever the true meaning, he had an undoubted influence on what became the 1967 'Summer of Love' centred in the Haight-Asbury district of San Francisco, but also celebrated around the world in an orgy of flowers, music and psychedelic drugs (see Figure 1.5). This outbreak of love and peace caused further alarm and consternation to a government that was fighting (and by now losing) the war in Vietnam, so it was inevitable that Nixon's *most dangerous man in America* would sooner or later fall foul of the law, and this he did on Boxing Day 1968. He was arrested in California allegedly in possession of marijuana, and despite his plea that the drugs had been planted, he was sent for trial. Whilst awaiting his court appearance he found time to announce his candidacy for the Governorship of California – against Ronald Reagan – and his campaign slogan was COME TOGETHER, JOIN THE PARTY. He also joined John Lennon and Yoko Ono at their well-publicised Montreal Bed-in in June 1969; but such activities were soon to be curtailed. At trial in January 1970 he was found guilty and sentenced to ten years in prison. Initially he was incarcerated in the maximum security prison at Chino in California, and there was talk of a transfer to the notoriously violent prison at Soledad, but Leary was fortunate enough to be given a psychology test that proved he would make a model prisoner so was transferred to the low-security prison at San Luis Obispo. What the prison psychologist did not know was that Leary had been responsible, at least in part, for the questions used in the personality test, so knew all of the right answers!

And that is where the story should end but Leary was such a larger-than-life character that it was almost inevitable he would escape. In an escapade that rivalled action seen in Indiana Jones movies, he escaped from San Luis Obispo prison on the evening of 12th September 1970. There are claimed to be at least 17 different versions of the escape story, mostly due to Leary. What seems to have happened is that he escaped by scaling a tree in the prison yard, clambering over the roof of the cell block and descending to freedom *via* a conveniently located telegraph pole. His wife Rosemary was waiting for him with a getaway car and with the additional help of the Weathermen – a revolutionary group bent on national disruption – he eventually flew out of New York on a false passport and into exile in Algeria. At this time Algeria had just emerged from a vicious colonial war of liberation with its French colonial masters, and the Algerian government was well disposed to other liberation groups from around the world, including the American Black Panther movement whose ambassador in Algeria was the infamous Eldridge Cleaver. The Learys sought refuge with Cleaver and, since they had brought with them a supply of high-quality LSD, the welcome was initially very warm. However, this warmth was soon compromised by Leary's enthusiasm for talking to the world's press including *Playboy* and *Rolling Stone* magazine. His claims for the sexually excitant properties of LSD, especially stories about the huge number of female orgasms that could be achieved on one dose, were widely reported. In addition, his outpourings about drug use caused embarrassment for the new revolutionary Islamic government. Soon the Learys were looking for a new place of exile and Switzerland provided their next place of refuge.

Interestingly, while in Switzerland he made arrangements through his lawyer to have lunch with Albert Hofmann, and the two duly met in a railway station snack bar in Lausanne. In his book *LSD: My Problem Child* Hofmann writes that he told Leary of his personal regret that promising scientific experiments with LSD and psilocybin had been overshadowed by the indiscriminate use of the drugs by the hippy culture. However, he went on to say that he was encouraged by Leary's very definite discrimination between the addictive drugs like morphine and heroin and the others (LSD, psilocybin and mescaline) that he felt were a force for good in the world. Hofmann also expressed concern that Leary was careless of his own safety, and this proved to be the case.

In January 1973 Leary's luck ran out when he was arrested on a trip to Afghanistan and repatriated to serve the rest of his sentence, though he was subsequently released early on licence in April 1976. The next 20 years were spent writing and giving lectures, though the glamour of LSD had now largely passed, primarily because its sheer potency meant that its effects were unpredictable. Other drugs like marijuana, cocaine and heroin were also much more easily available. Leary died in May 1996 and a year later some of his ashes were blasted into space along with those of Gene Roddenberry – creator of *Star Trek* – and others.

Although it is all too easy to exaggerate Leary's contributions to the history of LSD, the influence of this drug on the pop music of the late 1960s is usually held up as a very tangible example of its 'mind-expanding' properties. Compare, for example, the lyrics of most of the tracks of The Beatles' *Sgt. Pepper's Lonely Hearts Club Band* (*e.g. Lucy in the Sky with Diamonds*) or The Beach Boys' *Pet Sounds* or Bob Dylan's *Highway 61 Revisited* (especially *Desolation Row*) with the words of the songs on their earlier records, and the probable influence of LSD is apparent. John Lennon always claimed that the title *Lucy in the Sky with Diamonds* had nothing to do with LSD, but was inspired by a school painting produced by his young son Julian. Of the four Beatles, John appears to have been most influenced by Leary, and the latter was certainly present at the famous recording session of *Give Peace a Chance*, and appears in the video. The Beatles always claimed that they first experienced LSD in 1965 when someone 'spiked' their drinks at a party in London. In his definitive biography entitled *Shout*, Philip Norman records Cynthia Lennon's account of their nightmarish journey home, with John crying and banging his head against the car sides, and Pattie Boyd (George Harrison's wife) attempting to leave the moving vehicle in order to break windows in Regent Street. Nonetheless, Lennon reputedly took hundreds of LSD trips and this undoubtedly contributed to the growing instability within The Beatles.

In addition, a lot of so-called psychedelic art appeared during the 1960s but it is now generally accepted that painting while inebriated with LSD was not possible, since the rush of images was too great to be captured on canvas or paper. However, genuine works of psychedelic art could be inspired by the experience once it was at an end. More imaginative stories about the influence of LSD on Francis Crick's discovery (with James Watson) of the double helix and on Bill Gates' early work on computer programs have to be taken with a very large dose of scepticism.

The huge interest in LSD in the 1960s and 1970s, its restricted availability from the pharmaceutical industry and its relative ease of synthesis meant that there was a large cottage industry involved in making the drug illicitly. One particularly interesting example is worth relating since it was based almost wholly in the UK. The originator of this clandestine operation was the American Ron Stark, who was a member of the Brotherhood of Eternal Love, one of the many splinter groups that had grown out of the activities in Haight-Ashbury. It was this group who claimed to have funded Leary's escape from prison. Stark met and persuaded the young English chemistry graduate Richard Kemp to go into production. Kemp had obtained his B.Sc. at the University of Liverpool, and was studying for his Ph.D. (fluorine compounds and NMR spectroscopy) when he was persuaded, in 1970, to join Stark's illicit laboratory in the suburbs of Paris. Here he is alleged to have been involved in the synthesis of about a kilo of LSD. Having learnt his trade, Kemp then returned to England to live with his partner Christine Bott, who he had met at university while she was studying medicine. They set up a number of small-scale synthesis operations in various rented flats in and around London, and employed the one-time accountant Henry Barclay Todd to turn the synthetic LSD into tablet form (usually in the form of microdots on blotting paper) in a house in Seymour Road, Hampton Wick. These tablets were then passed on to various dealers, many of whom were students at Reading University, and were well placed to service the needs of those attending the various pop festivals in the Thames Valley area. A further participant, the American author David Solomon, travelled to Germany to buy ergo-tamine tartrate from the small company Dr Rentschler, and this was then stored in safety deposit boxes in Zurich and Geneva until Kemp collected it for transport back to England. He and Bott moved into a remote farmhouse at Blaencaron near Tregaron in Wales from around 1975, and another house at nearby Carno was rented for the actual chemical synthesis. The necessary purchases of chemicals and small bits of equipment were made by Todd (under a number of aliases) from British Drug Houses and from Hopkins and Williams. The paper records of these transactions were to prove vital as evidence in the eventual prosecution. From the initial batch of ergotamine it is claimed that they made around 1.7 kilos of LSD, which was enough to provide about 8.5 million doses of 200 micrograms each. These were sold as small tablets known as 'microdots' and were of exceptional purity. The operation went from strength to strength and by mid 1976 they were probably supplying about 95% of the UK's needs and about 50% of the worldwide requirements.

Most of the information about these clandestine operations comes from the book *Operation Julie* written jointly by Dick Lee (the Thames Valley police inspector in charge of the successful investigation) and the journalist Colin Pratt, and from the various TV documentaries that have been made about the operations. Kemp and Bott were only discovered after a series of misfortunes, most notably the impounding of Kemp's car after he was involved in a fatal accident. But even before this, in April 1975, Lee had been suspicious about the amount of LSD being seized at rock festivals in the Thames Valley region, though he was actually investigating the arrival of large amounts of cannabis in VW caravanettes, imported from Morocco into the Reading area. During an undercover operation, one of the local cannabis dealers revealed that he could obtain LSD for them, but then disappeared into a hippie commune near Tregaron before the police could extract further information. He was subsequently tracked down but would not reveal his sources, though by now Scotland Yard's drug squad had passed on to Lee the names of several people they were interested in, and these included Kemp and Bott. Fortuitously for the police investigation, the local Welsh police recalled the recent accident involving Kemp's red Range Rover, and the forensic team re-examined the vehicle. On this second search some scraps of paper were found which when pieced together revealed the word *hydrazine hydrate* – and someone had enough chemical knowledge to realise that this might be used in the LSD synthesis. In the early days of ergot research, hydrazine had been used to convert the complex ergot alkaloids into a mixture of lysergic acid and a residual peptide.

It took almost a year for Lee to persuade his superiors in the Thames Valley Police Force to fund further investigations, but finally in April 1976 the farmhouse at Blaencaron was put under police surveillance in a high-profile investigation code-named Operation Julie (named after one of the crime squad). Kemp was seen to spend several days at a time residing in the house at Carno, and the rest of the time with Bott at Blaencaron. In May, the police broke into the house at Carno while Kemp was away and, although there was little overt evidence of any major synthetic activities, they found two near-empty drums of methanol and a dead mole in one of the drains. Subsequent examination at the Government forensic laboratory at Aldermaston provided evidence of LSD intoxication in the mole, together with definite evidence of LSD in the drain effluent. This laboratory and its senior chemist – Neville Dunnett – were to have a key role in the analysis of material discovered in Wales, Hampton Wick and elsewhere. In addition, examination of records of water consumption at the Carno house showed that 500,000 gallons had been consumed over the previous 18 months, which was about ten times the amount consumed by a normal household and would support the existence of an illicit chemical operation.

Just as the investigation seemed to be moving towards a successful conclusion, Kemp abandoned the Carno laboratory and returned full time to domestic bliss with Christine Bott at Blaencaron. The distribution network was also proving difficult to identify until the police managed to infiltrate a group led by John McDonnell and William Lochhead in the summer of 1976. In a sting operation centred in Wiltshire, the undercover agents managed to buy LSD tablets which were matched to others believed to have originated from the Welsh operation. Then, in November, Kemp began to extend the plumbing in a newly rented cottage at Esgair Wen about four miles from Lampeter, and this work included a new drain system to lead effluent into a nearby stream. Obviously the synthetic operation was about to restart.

Finally, after a winter of further observation and evidence gathering, the house in Seymour Road was raided on 25th March 1977. Todd's tableting laboratory was discovered, complete with appropriate chemicals, drying trays, some LSD tablets, around 250 mg of pure LSD and about £20,000 hidden in packets of Alpen cereal. Todd and several accomplices were arrested in the house. On the next day, the cottages at Esgair Wen and Blaencaron were raided and Bott and Kemp were arrested. Under interrogation, Christine Bott revealed the location of some of the LSD that had been produced, together with the existence of a safety deposit box in a Zurich bank. Later excavation under the kitchen floor tiles in the Blaencaron cottage revealed around 1.3 kg of pure LSD worth about £65 million on the street. At the subsequent trial, 17 persons were found guilty of various offences. Kemp and Todd both received sentences of 13 years, while Bott received one of 9 years, and 14 others received sentences ranging from 2–11 years. Although more of the LSD tablets were recovered from caches in woods in Berkshire, most of the money that was undoubtedly earned by the participants has never been recovered.

The mode of action of LSD is still poorly understood and its ability to cause changes to the senses, for example how it allows people to 'hear' colours and 'smell' sounds, is almost impossible to comprehend. It can also precipitate flashbacks many weeks after it has been ingested and there have always been concerns that frequent use could lead to long-term psychosis. What is known is that the main pathway by which it exerts its effects is *via* interference with the normal functioning of the neurotransmitter serotonin (5-hydroxytryptamine – 5-HT) within the brain. In particular, it competes with serotonin for 5-HT receptors, especially in the hypothalamus and brain stem, though this competition is far from uniform in different parts of the brain. Very recently, Stuart Sealfon and his colleagues in the Mount Sinai School of

Medicine, New York, have shown that LSD and the other hallucinogens like mescaline and psilocybin, interact with a specific subclass of cortical neurons which carry 5-HT subclass 2A receptors and lead to activation of unique signalling pathways. This activation leads to the observed neurophysiological responses, though we are still far from understanding how the hallucinations are produced.

In May 2008, approval was given for a Swiss study that will explore the use of LSD to ameliorate anxiety in terminally ill cancer patients, and this study received FDA support in September 2008. Whatever the future holds for LSD research, there is no question that discovery of its interactions with the 5-HT system contributed to other research that led to the discovery and exploitation of anti-depressants like Prozac (fluoxetine) and Zoloft (sertraline), and also the migraine drugs like Imigran (sumatriptan). Both of these conditions – depression and migraine – invlove disturbance in 5-HT neurotransmission.

In the UK, LSD is categorised as a Class A drug in the same group as heroin and cocaine so it is illegal to supply it, to buy it, and to use it, and any of these crimes can attract a prison sentence of up to 10 years. The other ergot alkaloids no longer have widespread use within a clinical setting and thankfully they are only very rarely responsible for outbreaks of ergotism. The story of the ergot alkaloids is thus essentially complete, though the three larger-than-life characters R. Gordon Wasson, Timothy Leary and Albert Hofmann will be remembered for a long time to come. In an interview with Leary just before he died, he was asked by a reporter what he thought of Nixon's claim that he was the most dangerous man in America. Leary replied: *Yes it's true – I have America surrounded*, which provides an amusing epitaph for this interesting man. But the final words must be those of Albert Hofmann, who died in May 2008 aged 102:

During the first years after its discovery, LSD brought me the same happiness and gratification that any pharmaceutical chemist would feel on learning that a substance he or she produced might possibly develop into a valuable medicament. The more its use as an inebriant was disseminated, bringing an upsurge in the number of untoward incidents caused by careless, medically unsupervised use, the more LSD became a problem child for me.

CHAPTER 2

Opiates from Opium to Heroin

Yea, slimy things did crawl with legs
Upon the slimy sea.
About, about, in reel and rout
The death fires danced at night;
The water, like witch's oil,
Burnt green, and blue and white.

This is how the English poet Samuel Taylor Coleridge described the horrors seen by the crew in his *The Rime of the Ancient Mariner*, and while the poem was written before he became addicted to opium, there is little doubt that he wrote the poem while treating himself with opium for his dysentery. Such influences of opium on artistic output much postdate its medical uses, and these uses were probably discovered by the ancient Sumerians as long ago as 3400 BC. Once again there is evidence from the ideograms on Sumerian tablets where *hul* and *gil* have been translated as the expression 'joy plant'. Other tablets have revealed the method of collection of crude opium that involved scoring the unripe seed capsules of the poppy plant with a knife, to allow a milky exudate to 'bleed' from the capsule (see Figure 2.1), then returning the next day to scrape off the congealed latex. Often the harvesters would walk though the fields wearing leather aprons to which the latex adhered. The exudate was then

Turn On and Tune In: Psychedelics, Narcotics and Euphoriants
By John Mann
© John Mann 2009
Published by the Royal Society of Chemistry, www.rsc.org

Figure 2.1   An unripe seed capsule of the opium poppy scored so as to release the crude opium exudate. © Henry F. Oakeley, Garden of the Royal College of Physicians.

usually treated with boiling water in which it dissolved, impurities were filtered off, then the water evaporated to leave a brown, mobile liquid which dried to leave a thick brown paste of crude opium. These methods have changed surprisingly little over the millennia. Of the numerous species of poppy only *Papaver somniferum* (and to a lesser extent *Papaver setigerum*) contains significant quantities of opium alkaloids, with morphine comprising up to 20% by weight of crude opium (with codeine up to 2.5% and thebaine up to 2%). Although there is ample evidence of seeds of both plants in Neolithic sites, only *Papaver somniferum* was domesticated to provide the plant that we call the opium poppy.

None of these early references to the opium poppy mention recreational use, and right up until mediaeval times there was an almost complete emphasis on medical uses, though these did include mixtures of opium and various poisons (especially hemlock) as agents for euthanasia. It was probably the ancient Egyptians who first popularised the medical uses of opium, and the Ebers papyrus, from about 1500 BC, contains several hundred preparations for the treatment of myriad conditions including hair loss (no likely efficacy) and childhood colic (excellent chance of success). Indeed opium produced in Thebes became famous from around 1300 BC and was exported to numerous countries that bordered the Mediterranean Sea. The growing importance of opium from around this time is clear from its association with deities in many societies like the Minoan goddess (of narcotics?) depicted with her garland of poppies, the minor deities associated with sleep (Greek-Hypnos and Roman-Somnus) and of death (Thanatos) usually shown with poppy wreaths. Most references to other medicinal uses of opium come to us from ancient Greek texts, and Homer's *nepenthe* (from ne = not, penthos = grief) was probably a mixture of opium and wine. In the *Odyssey* he describes its use as an aid to forgetting all the trials and tribulations of the Trojan war:

> *Helen, daughter of Zeus, poured a drug, nepenthe, into the wine they were drinking which made them forget all evil. Those who drank of the mixture did not shed a tear all day long, even if their mother or father had died, even if a brother or beloved son was killed before their eyes by the weapons of the enemy.*

The great Greek physicians like Dioscorides, Theophrastos and Hippocrates all recognised that 'poppy juice' induced sleep and could reduce pain. However, although all of them did use what the Greeks termed ὄπιον they also appear to have used extracts from a variety of other species of poppy which would not have had the same powers associated with *Papaver somniferum*. Later, Greek medicine was the main influence in the Roman Empire and probably the greatest advocate of opium was the Greek physician Galen (second century AD). He claimed the poppy extract as a panacea which he christened *mithridatium*, and prescribed it for just about every condition from colic and headache (where there was real efficacy) to leprosy and menstrual problems (doubtful efficacy); but he also made a genuine contribution to the study of the toxicity of opium and in particular he noted the induction of tolerance such that the patient needed to take larger and larger doses in order to have the same effect. He reported that the Emperor Marcus Aurelius used a mixture of opium and honey to help him sleep. As usual the Romans were not slow to take advantage of new opportunities to dispatch one another, and it is possible that the Emperor Claudius' fourth wife Agrippina poisoned her stepson Britannicus with opium in AD 55 so that her son Nero would have no rival to his claim to become Emperor. If true, an early example of the malevolent influence of opium on history! In addition, the Romans were the first to take seriously the recreational possibilities of opium, though there is little reliable record of excesses like those observed later in China.

As with all empires the Roman influence waned, and from around the fourth century the Arabs became the dominant force in the eastern Mediterranean and the Middle East. Their influence on medicine and learning was unrivalled in the region for nearly a thousand years from AD 500–1500, and many of their contributions to architecture still provide testimony to this golden age. Within their great library in Alexandria they maintained copies of all of the great Greek and Roman texts on medicine and thus ensured that the medical lore of giants like Galen and Dioscorides would survive right up until the present day. In particular, the mammoth five-volume *De Materia Medica* of Dioscorides written between AD 60 and 78, and the 20 volumes of Galen's *Opera Omnia* (which appeared around AD 170), formed the basis of most prescribing right through to Elizabethan times. The Arab medical experts embraced opium as a treatment for pain and stomach disorders, and as their empire spread westwards into Spain and eastwards as far as India, they took opium with them. In fact they were probably the first to introduce the plant extract into China. Of the Arab physicians Abu Bakr Al-Razi or Rhazes (AD 865–923) and Abdallah Ibn Sina or Avicenna (AD 980–1037) were the most influential. The former was both a medical expert and an alchemist who extolled the virtues of opium in anaesthesia, and was also experienced in the science of distillation. Avicenna managed to combine an adventurous lifestyle with a prolific output as an author of philosophical, medical and herbal texts, many of which were still in use centuries later. His *Canon of Medicine* (translated into Latin by Gerardo de Cremona around AD 1140) was the most influential, and of the 760 drugs mentioned it highlighted the value of opium as a narcotic. It was this book and the works of Galen that were ceremonially burned in 1526 by that most infamous of sixteenth century physicians Philippus Aureolus Theophrastus Bombastus von Hohenheim – usually known as Paracelsus. Paracelsus was born near Zurich in 1493 and after training periods in Switzerland, Germany and Austria received his doctorate in medicine from the University of Ferrara. After several years of travelling, he was appointed Professor of Medicine in Basel in

1526, promptly vilifying the plant-based drugs of Galen and Avicenna and publicly burning their books. He then embarked on a crusade that emphasised the salts of arsenic, antimony, tin and mercury for treatment of all manner of ailments. He was also not above using magic, necromancy and astrology, and is reputed to have introduced the use of *laudanum* to mankind. He claimed *I possess a secret remedy which I call laudanum and which is superior to all other heroic remedies.* That he invented laudanum is almost certainly untrue, but his recipe for this is claimed to have included about 25% of opium mixed in with henbane, amber, musk and various more unusual ingredients including bezoar stone – in reality an accretion from cow's intestines – and crushed unicorn horn – in reality probably crushed sea shells. Whatever its actual composition, laudanum became synonymous with the various preparations of opium that were dispensed, most of which later involved dissolution of the poppy extract in wine or purer forms of alcohol.

Another populariser of opium was the Spanish physician Cristobal de Acosta, who produced, in 1582, a treatise on medicines from the East Indies. His work was widely regarded not least because he recommended opium as an enhancer of sexual performance and particularly extolled its virtues in the prevention of premature ejaculation. Other famous herbalists enthused about the benefits of opium, and John Gerard, undoubtedly the most important Elizabethan adviser on drugs, provided a very apposite description in his *Great Herball* (1597):

> *It mitigateth all kinde of paines, but it leaveth behind it oftentimes a mischief worse than the disease itselfe.*

But the greatest populariser of opium was Thomas Sydenham, born (or at least christened) in Wynford Eagle, Dorset, in 1624. His studies at Oxford University were interrupted by the Civil War, and he participated in four years of bloody conflict with the Parliamentary forces from 1643–1646 before returning to study medicine. He received his Bachelor of Medicine from Oxford in 1648, and then practised as a physician first in Oxford and later in London. Although he is most often associated with the introduction of the bark of the *Cinchona* tree (and hence quinine) as a treatment for malaria, which was endemic in many parts of Britain, he was also fulsome in his praise for opium:

> *Among the remedies which it has pleased the Almighty God to give to man to relieve his sufferings, none is so universal and so efficacious as opium.*

And in the first volume of his *Works*, first published in Latin in 1676 (with a translation into English in 1742 by John Swan), he summarised the utility of opium:

> *So necessary an instrument is opium in the hand of a skilful man, that medicine would be a cripple without it; and whosoever understands it well, will do more with it alone than he could well hope to do with any single medicine. To know it only as a means of procuring sleep, or of allaying pain, or of checking diarrhoea, is to know it only by halves.*

His preparations were based upon opium and saffron in the ratio of 2 : 1 combined with a pinch of cinnamon and cloves (to mask the bitter taste), and all dissolved in canary wine.

Sydenham was friendly with several of the soon-to-be-famous Oxford crowd, and of these it is recorded that Christopher Wren and Robert Boyle experimented in 1656 with the new technique of injection using a hollow quill. They reported that a dog became extremely stupefied after an injection of a warm solution of opium.

The first major work in English that concentrated on the uses of opium was provided by the Welsh physician John Jones in 1700 in his book *Mysteries of Opium Reveal'd*. He assured his readers that:

> *It takes away grief, fear, anxieties, peevishness, fretfulness . . . and causes a great promptitude to venery and erections.*

But he did warn of the severe suffering of withdrawal from opium addiction:

> *Great and even intolerable distresses, anxieties and depressions of spirits which in a few days commonly end in a most miserable death.*

One of Thomas Sydenham's pupils, Thomas Dover, invented the very popular Dover's Powder, which comprised opium, liquorice, ipecacuanha (usually an emetic), saltpetre and vitriolated tartar all to be mixed and taken in white wine. In his youth he travelled widely and was for a while a privateer. It is even claimed that in 1709 he was the saviour of Alexander Selkirk, who had been marooned on the island of Juan Fernandez off the coast of Venezuela, and who later became the model for Robinson Crusoe. But even the popularity of Dover's Powder paled in contrast to Godfrey's cordial (a mixture of opium, spices and treacle in water), which was being sold at the rate of several gallons per week by some pharmacies in the big industrial cities like Manchester and

Nottingham during the eighteenth and nineteenth centuries. Other popular preparations included Mrs Winslow's Soothing Syrup, Mother Bailey's Quieting Syrup, Daffy's Elixir and Atkinson's Infants' Preservative.

The reasons for these quite extraordinary sale volumes are not hard to discern when one appreciates that laudanum sold for one penny per ounce which amounted to half the price of a pint of beer. Opium was used for a whole range of ailments from coughs and colds to rheumatism, and a remedy for piles comprised laudanum and egg yolk in an ointment. It was also much used for controlling the symptoms of the very prevalent dysentery and cholera, but also to keep babies quiet. In order to make ends meet, many wives worked in the mills and other factories of the big industrial centres and had to pay for the care of their small children. The babyminders would often reduce the cost of their services if the parents provided them with so-called 'quietening mixture', and the constant dosing of the children with opium often led to malnutrition due to an associated disinclination to eat. In a report to the Privy Council in 1862 (actually the fourth Annual Report of the Medical Officers of Health), a Dr Edward Greenhow suggested that the high rates of infant mortality in these big cities was as much to do with malnutrition as it was to dysentery and infections, though he might have added that opium was also implicated in accidental or deliberate overdosing of children.

The eating of opium was also common in certain parts of the country, especially the Fens – north from Cambridge to the Wash. This was allegedly because rheumatism was prevalent in this part of Britain and the many agricultural workers used it to ease their pains, but probably also to alleviate the misery of their lives in what was then a damp and dreary part of Britain. The overall level of opium dependence can only be guessed at, but was probably quite high.

The popularity of opium in eighteenth and nineteenth century Britain supported a huge supply and distribution network, and this had its origins in the highly profitable activities of the East India Company. This venerable body was originally heavily involved in the supply of opium to China and thus responsible, in part, for the abuses of opium in that country. As already mentioned, Arab traders did supply opium to the Chinese but it was almost certainly already present in China from as early as the first century AD. There were imports from Burma together with home-grown opium poppies, and most of the opium was used for medicinal purposes prior to the sixteenth century, though there was some recreational use through eating opium with its bitter taste masked by the addition of various spices like nutmeg, cardamom and cinnamon.

This route of administration was even more widespread in India and Turkey. The smoking of opium was unknown in China until smoking of tobacco became popular in the Far East. This happened in the late sixteenth century as the Portuguese, Spanish and Dutch traders took advantage of the earlier discoveries in the Americas to introduce the tobacco habit into their colonies in India, Indonesia and the Philippines, and subsequently into China and Japan. The Chinese embraced this new habit with enthusiasm and tobacco smoking became a real social problem, so much so that it was banned for a while during the early seventeenth century. It was the vacuum produced by this ban that opened the way for the introduction of opium smoking – firstly as a mixture with tobacco which had become popular in the Dutch colony of Java – but then in pure form. Opium shops and dens quickly appeared in most towns and these catered for the growing number of addicts. The smoker inhaled the fumes from a globule of molten opium held in the bowl of a bamboo pipe that was heated in a flame, and these fumes were then held in the lungs for as long as possible before expulsion through the nostrils. Serious smokers would take as many as four pipes in quick succession before succumbing to a deep and troubled sleep (narcosis) for periods lasting from 15 minutes to several hours, and would then awaken without apparent ill effects.

The trade in opium was initially dominated by the Portuguese *via* their settlements in Goa, then the Dutch took over using their colonies in Java, and finally the British dominated the trade using the agents and ships of the British East India Company. This hugely successful trading operation had been established in 1600 through a charter issued by Elizabeth I, and awarded (by the British Government) exclusive trading rights with all lands *beyond the Cape of Good Hope and the Magellan Straits*. The importance of the opium trade to the EIC was forcefully demonstrated when the Governor General Warren Hastings established a state monopoly in November 1773. This required all Indian poppy growers to sell their produce to the company, yet to preserve the Indian labour force from this addictive and debilitating product, all of the product had to be exported. Warren Hastings famously stated:

Opium is not a necessity of life but a pernicious article of luxury, which ought not be permitted except for the purposes of foreign commerce only, and which the wisdom of the Government should carefully restrain from internal consumption.

Such two-faced behaviour was unfortunately all too common in dealings between officials of the EIC and its colonial partners and employees.

Imports of opium into China grew from a modest 200 chests in 1730 to more than 2000 chests (perhaps 125 tons) by the end of the eighteenth century and this ensured that the level of addiction in the population grew alarmingly – to about 1% of the population by the 1760s. Most of the importation took place *via* Canton at the mouth of the Pearl River (see Figure 2.2) and in 1837 there were more than 150 EIC agents in the city. Eventually the opium trade represented around 40% of its total exports, but was offset by the negative balance of trade occasioned by the large tonnage of exports of silk and tea that left China for Britain, and this trade imbalance had to be paid for from the silver reserves of the company. The situation gradually improved for the EIC, and by 1840 more than 40,000 chests of opium were imported at a cost to the Chinese exchequer of five million silver dollars.

The Chinese authorities, alarmed by this outflow of bullion and also by the levels of addiction, made efforts to halt the trade. The Emperor Kia King banned importation in 1799 but this had little effect due to organised smuggling involving the various importers and corrupt Chinese officials. However, this did cause aggravation for the EIC and its agents in Canton since the company could not directly contravene the ban. They circumvented these obstacles by selling their opium to so-called Free Traders in Calcutta, who were mainly English and

Figure 2.2 A view of opium traders on the Pearl River. © Jupiter images.

Scottish entrepreneurs, and these organised shipment onwards to China. Some idea of the success of these operations can be appreciated from figures showing that in 1825 more than 10,000 chests were imported *via* Lintin, an island off Macao and 80 miles down the Pearl River estuary from Canton. Inward bound ships unloaded their cargoes onto floating hulks that were mainly ex-merchant vessels whose masts had been removed and were essentially fortified opium warehouses. The situation was exacerbated for the EIC when the British Government revoked its monopoly of trade with China by means of the India Act of 1833, and as a direct result the company withdrew its staff from Canton. There was also competition from a growing number of British and American owners of fast opium clippers who carried opium from India to China.

Relations between the Chinese authorities and these traders gradually worsened and major ferment was caused when the Emperor Tao-Kuang sent a newly empowered commissioner Lin Tse-hsu to Canton in March 1839 with the authority to seize and destroy all of the opium stocks. Not surprisingly the European traders initially refused to comply with his edicts. However, they then received instructions from Lord Palmerston's chief representative, Captain Charles Elliot, a former Royal Navy sea captain, to comply and duly handed over in excess of 20,000 chests of opium, much of it belonging to the British firm of Jardine, Matheson and Co. There was method in this apparent madness since the merchants were overstocked with opium and the price had fallen, and in addition they had received a promise from Queen Victoria's Government that they would be indemnified for their losses. Lin Tse-hsu now found himself with a major problem of destruction of this contraband. He eventually solved this by stirring the opium with a mixture of salt and lime in water in large trenches, and once the opium had been broken down the detritus was flushed away into the Pearl River.

Despite this setback, the traders were soon back in business importing opium *via* ports further up the coast, and this precipitated a brief naval skirmish off Hong Kong in September 1839 between the British merchant fleet and the blockading Chinese junks. Further naval actions followed and an expeditionary force of 10,000 men plus 18 warships were rushed from the colony of Ceylon in April 1840, and the First Opium War commenced in earnest. This was a very uneven conflict since there was essentially no Chinese navy while the British merchant and naval fleets were the strongest in the world. The war dragged on for two years with a series of small but bloody battles resulting in the capture of Shanghai and the ports of Ningbo, Zhapu and Zhenjiang. The naval forces sailed up the Yangtze River reaching Nanking (now Nanjing) in August 1842.

The Chinese now sued for peace and a treaty was signed on board *HMS Cornwallis* in August with the Chinese agreeing to cede the barren island of Hong Kong to the British in perpetuity with a payment of 21 million silver dollars as recompense for the opium destroyed. For the Chinese this was not only a financial disaster but a practical disaster as well, since opium imports quickly rose to pre-war levels almost immediately and continued to grow to match demand. Any satisfaction felt by the British War Office was tempered by a growing feeling in Victorian society that the opium trade with China was immoral. The parliamentary view at the time is probably best summed up by the statement of the future Earl of Shaftesbury when he claimed:

I am fully convinced that for this country to encourage this nefarious traffic is bad, perhaps worse than encouraging the slave trade.

Despite these misgivings the trade with China, especially in opium, continued to expand, and by 1850 75% of Indian opium passed through the colony of Hong Kong. Following the peace treaty of 1842, Hong Kong had developed rapidly with all the normal essentials for colonial life including fine houses, shops and brothels, as well as a fine bonded warehouse for the main opium exporter, Jardine, Matheson and Co. The Americans and the French were also now seeking their share of this lucrative trade, and Britain retaliated by seeking 'most-favoured-nation' status from the Chinese in 1854 with demands that the Chinese open all their ports for duty-free importation of British goods. The Chinese refused and, following several incidents including an attempt to poison much of the European population in Hong Kong, joint military action by the British and French was initiated in what became known as the Second Opium War (1857–1860). This was a much more bloody conflict than the first war and at its height involved nearly 18,000 allied troops and a naval squadron numbering 173 vessels. It culminated in the capture of Beijing in October 1860 after a pitched battle that matched elite (but doomed) Mongolian cavalry against allied cannons. The allies then looted much of the valuable artwork from the Old Summer Palace before destroying the buildings, but stopped short of destroying the Forbidden City for fear that such an act of supreme vandalism would jeopardise any hope of bringing the Chinese to the peace table. In the event, the Chinese were forced to allow further trading opportunities and also to officially legalise the opium trade, and this lasted right through until 1906 when a decree was issued that demanded the immediate closure of all opium dens and the cessation of smoking. Even as late as 1930, there were in excess of 6000 licensed opium dens in Southeast Asia, 3000 of these in French Indochina. In the meantime, Chinese migrants had taken their habit with them into the rest of Southeast Asia, Australia and across the Pacific to the USA and South America. Tens of thousands of Chinese migrants entered the USA *via* San Francisco and worked in the Chinese laundries, on the railroads and in the goldfields of California.

Migration into Europe was less common though opium dens did exist in most of the large cities (see Figure 2.3) and, as in the USA, these were considered to be centres of vice and prostitution, though seduction was usually the last thing on the minds of the most dedicated opium smokers. It was the artists, poets and writers of the so-called Romantic era (around 1775 to 1835) who adopted opium as their drug of choice. Perhaps most famous of these was Thomas Penson de Quincey (born in Manchester in 1785, see Figure 2.4) with his book *Confessions of an English Opium-Eater* published in 1822 – it had first appeared in *The London Magazine* in two parts in 1821. He seems to have first used the drug in 1804 as a treatment for toothache and

LONDON SKETCHES—AN OPIUM DEN AT THE EAST END

Figure 2.3 An East End opium den in the eighteenth century. © Wellcome Images.

Figure 2.4 A signed portrait of Thomas de Quincey. © Bettmann/Corbis.

facial neuralgia, and the pain relief was but a prelude to a new pleasure. In his book he noted:

> *That my pains had vanished was now a trifle in my eyes – this negative effect was swallowed up in the immensity of those positive effects which had opened before me – in the abyss of divine enjoyment thus suddenly revealed. Here was a panacea . . . for all human woes; here was the secret of happiness.*

He began taking laudanum on a regular basis to counteract the pain of a gastric ulcer, but found that it also enhanced his enjoyment of weekly trips to the opera. In publicising how the drug enhanced his *sensual pleasure* of music, he undoubtedly encouraged others to experiment with the drug, just as later generations would use LSD or marijuana for enhancement of sensual experiences. The section of his book entitled *The Pleasures of Opium* contained very strong endorsements for the advantages of opium use compared with alcohol consumption:

> *Wine disorders the mental faculties, opium on the contrary introduces among them the most exquisite order, legislation, and harmony. Wine robs a man of his self-possession: opium greatly invigorates it.*

However, his experimentation was soon out of hand and by 1815 he was taking 8000 drops of laudanum each day and had descended into penury with nightmarish visions of being *buried, for a 1000 years, in stone coffins, with mummies and sphinxes... I was kissed, with cancerous kisses, by crocodiles; and laid, confounded with all unutterable slimy things amongst reeds and Nilotic mud.* It was to revive his fortunes that he wrote the *Confessions* and the book was popular not least because it captured the imagination of a literature-loving middle class who revelled in tales of roguish behaviour and rebellion. Despite his reputation as a reprobate, de Quincey appears to have been a sensitive and generous man, who became very friendly with both William and Dorothy Wordsworth, and the Coleridges. At one point he lent money to Coleridge, and became particularly fond of the Wordsworth children. His familiarity with their lives and literary works allowed him to write critical accounts of these important poets which helped enhance their reputations. There were several editions of his book, which was eventually twice its original length, and de Quincey was reviled and admired in equal measure for his admissions. The proceeds from the book certainly allowed him to lead a reasonably happy life, and eventually he managed to reduce his consumption of opium from 320 grains a day (about 8000 drops) to 12 grains (around 300 drops). He clearly suffered severe withdrawal effects during this process and wrote of *the torments of a man passing out of one mode of existence into another.* He died in 1859 at the age of 74 years, which was a good age considering his abuses. His near contemporary, the poet Samuel Taylor Coleridge, was not so lucky.

Like de Quincey, Coleridge began taking laudanum as a treatment for neuralgia and toothache, but also to alleviate the symptoms of dysentery. His consumption of the drug increased dramatically in 1802 as he treated himself for a back condition and rheumatism, and he became increasingly miserable from what he described as *the worst and most degrading of slaveries.* He was certainly not helped by the time he spent with the notorious Thomas Beddoes of Bristol, who experimented in his Pneumatic Institute with opium, henbane, cannabis and even nitrous oxide. On innumerable occasions Coleridge tried to break himself of the habit and was, for the period 1816 to 1828, a tenant of the physician James Gillman in Highgate, London, who tried to wean him off laudanum, little realising that Coleridge was being supplied secretly by a local druggist. At times he was consuming as much as two pints (perhaps 20,000 drops) of laudanum each week. And yet he managed to produce some stunning poetry including his best-known works *The Rime of the Ancient Mariner* and *Kubla Khan*, and although the latter was often considered a classic example of a work inspired by opium, this has now been discounted. Coleridge's plight was perhaps best summarised by the author Sir Walter Scott when he claimed that Coleridge was *a man of genius struggling with bad habits*. When he died in 1834 (aged 62) he left a written plea that others might learn from his mistakes:

After my death, I earnestly entreat that a full and unqualified narrative of my wretchedness, and of its guilty cause, may be made public, that at least some little good may be affected by the direful example.

The poets Byron, Shelley, Keats and Southey were all moderate consumers of opium, though only Keats makes much mention of it in *Ode to a Nightingale*:

My heart aches, and a drowsy numbness pains
My sense, as though of hemlock I had drunk,
Or emptied some dull opiate to the drains.

And in *The Eve of St. Agnes*:

In sort of wakeful swoon, perplex'd she lay,
Until the poppied warmth of sleep oppress'd
Her sooth'd limbs, and soul fatigued away.

Elizabeth Barrett, later to become the wife of the poet Robert Browning, appears to have become a regular user of opium at the age of 15 as the result of a painful spinal condition. For her it was an *elixir because the tranquilising power has been so great* and later she claimed:

I think better of sleep than I ever did, now that she will not come near me except in a red hood of poppies.

Her condition was never completely diagnosed and the pain was often so great that she complained *opium, opium – night after night – and some nights even opium won't do.* She nonetheless produced some very well-regarded poetry, some of which bears the hallmarks of opium intoxication.

Across the English Channel, opium use was no less common and the French poet Charles Baudelaire was first induced to take the drug as a treatment for the syphilis he acquired as a young man – *when I was very young I got poxed* – he is reputed to have told his mother. In his poem *La Chambre Double* he compares the lures and perils of women and opium:

Here in this world, cramped but so disgusting,
Only one familiar object cheers me:
The phial of laudanum;
An old and terrifying friend;
Like all women, alas,
Prolific in caresses and betrayals.

As for writers of prose, Wilkie Collins was possibly the most conspicuous of the opium addicts. His most famous novel, *The Moonstone*, was probably composed almost wholly under the influence of the drug and it was his secretary who wrote the actual words in 1868 as he dictated to her. Certainly the plot relies very heavily on the theft of a large Hindu diamond carried out while the thief is under the influence of opium. Collins made no secret of his addiction and carried a hip flask of laudanum wherever he went, describing himself as someone for whom *laudanum – divine laudanum – was his only friend.* He was also one of the first of the literati to take injections of morphine, ostensibly to break him of the habit of drinking laudanum. In a letter to a friend he reported:

My doctor is trying to break me of the habit of drinking laudanum. I am stabbed every night at ten with a sharp-pointed syringe which injects morphia under my skin – and gets me a good night's rest without any of the drawbacks of taking opium internally.

Despite this opiate abuse, which probably commenced as a treatment for his rheumatism and gout in 1862, he nonetheless survived until 1889 and died aged 65 years.

One nineteenth century writer and poet who is invariably mentioned whenever drugs and literature are linked is Aleister Crowley (born in

1875 in Leamington Spa). He is probably most remembered for his interests in the occult and mysticism which he claimed were initiated following a 'visitation' by the goddess Nuit while he was on vacation in Egypt in 1904. This led him to expound the philosophy of the Law of Thelema and to his penning a sequence of books about this religion. He was also the author of books with titles like *Magick without Tears* and *Diary of a Drug Fiend*, and some poetry of which *White Stains* published in 1898 is probably the most infamous being a collection of pornographic verse which was immediately banned. He fuelled a libidinous and bisexual lifestyle with alcohol, opium, heroin, cocaine, hashish, mescaline and even ether, and had the dubious honour of being immortalised in Somerset Maugham's 1908 novel *The Magician*, and appearing amongst that famous collection of persons on The Beatles' *Sgt. Pepper* LP sleeve. Despite his dissolute lifestyle, he survived to the ripe old age of 72 years.

Other nineteenth century novelists probably took opium occasionally for pain and described their experiences of seeing the darker side of life in the ghettos of, for example, London in the cases of Charles Dickens (*The Mystery of Edwin Drood*) and Oscar Wilde (*The Picture of Dorian Gray*).

Dorian winced, and looked round at the grotesque things that lay in such fantastic postures on the ragged mattresses. The twisted limbs, the gaping mouths, the staring lustreless eyes fascinated him. He knew in what strange heavens they were suffering, and what dull hells were teaching them the secret of some new joy.

The Royal Family was also familiar with opium and other substances, as the records of the local pharmacy near Balmoral have recently revealed. For George IV substance abuse became a way of life, and his binge drinking of opium and cherry brandy mixtures together with his gluttony hastened his descent into invalid status, though he somehow managed to survive until age 68.

One further supposed attribute of opium was its ability to control what the leading physician Sir Almroth Wright described in 1912 as a woman's *periodically recurring phases of hypersensitiveness, unreasonableness, and loss of the sense of proportion.* These problems were because *the physiology and psychology of women is full of difficulties.* He expounded these views initially in a letter to *The Times* in March 1912 and then in expanded form in a treatise entitled *The Unexpurgated Case Against Woman Suffrage* published in 1913. Similarly misogynistic views and assertions appeared in the medical textbooks of the time and those of the gynaecologist Dr Henry McNaughton Jones are particularly extreme. His textbook *Diseases of Women* went through nine editions, and in this he revealed the particular challenges of the unmarried woman who he believed to be generally neurotic and who harboured *erotic thoughts, desires and practices that still further enervate her nervous systems and enfeeble her central control.* Married women were no less problematic with *their congestive dysmenorrhoea and overalgia, her uterus may be as flabby as her brain, and her ovary be as fertile in aches as her imagination is in fanciful illusions.* These conditions might be treated with opium though were, he believed, more often caused by dependence on opium. Certainly there was a large number of middle-class Victorian wives for whom opium provided a release from their boredom and frustration. For many Victorians, both male and female, opium use was as common as their use of tobacco.

These literary experiences and other excesses were made possible by the almost unlimited access to poisons and potions during Victorian times. Opium use was not regulated in Britain until the Poisons and Pharmacy Act of 1868, and supplies of the drug poured in from Turkey and India. There are records from 1830 showing that around 190,000 pounds of opium was imported into England (97% from Turkey), while by 1900 the amount had risen to around 620,000 pounds (74% from Turkey, 12% from India and lesser amounts from Persia and France). Major wholesalers included William Allen and Company of Plough Lane, London (later to become part of Allen & Hanburys and then Glaxo), and Thomas Morson and Sons of Fleet Market, London. These supplied the provincial apothecaries and thence the corner shops whose owners had little knowledge of the drug they were selling – often to children who had been sent on an errand to buy *a penny stick of opium.*

With the establishment of the Pharmaceutical Society of Great Britain in 1841 (allegedly by a group of London chemists and druggists who met in the Crown and Anchor tavern in the Strand), a new view of opium use began to emerge that questioned its safety. There were special concerns with regard to accidental overdose, suicide (throat cutting and hanging were the favoured methods), murder and the ever-present problem of adulteration – Turkish opium regularly contained quantities of cow dung, poppy capsule material and even gravel. Attempts were made to weed out charlatans and the Pharmaceutical Society Bill of 1851 stated that:

It should not be lawful for any person to carry on the business of a chemist and druggist . . . unless such a person shall be a pharmaceutical chemist.

The Poisons and Pharmacy Act of 1868 led to the introduction of formal registration for pharmacists after special examinations. In addition, opium was officially identified as a potential poison which required appropriate labelling, though patent medicines were excluded from the bill and this loophole allowed the famous Dr Collis Browne's Chlorodyne to remain on sale. This mixture of chloroform, opium and (usually) cannabis, launched in 1856, was in use for everything from cholera to flatulence with some people consuming several pints each week. Certainly the dangers of addiction were well understood and these were clearly enunciated in Louis Lewin's *Phantastica: Narcotic and Stimulating Drugs*:

> *The drug* (opium) *must be injected more frequently and in larger quantities, the chain of slavery becomes shorter and tugs at the morphinist ... the brain cells grow restive, demand satisfaction, shriek, and revenge themselves by producing pain if they are not satisfied quickly enough.*

Against this background of widespread use and abuse, most of the groundbreaking scientific studies on the chemical constituents of opium took place during the nineteenth century. The French pharmacist Armand Séguin was probably the first to isolate morphine in crystalline form in 1804, and described his work to the Institut de France in Paris under the title *Sur l'Opium*; but he made no written report before 1814. So it was the German Friedrich Wilhelm Sertürner, a young pharmacist's assistant with no formal scientific training, who is usually given the credit for the first isolation of morphine. He reported his discovery, which he called 'morphium' after the Greek god of dreams Morpheus, in a relatively obscure journal – the *Journal der Pharmacie* – in 1806. He and three associates went on to experiment with this new crystalline substance and he evidently experienced many of the side-effects of the large doses he is reputed to have taken. He reported these new results in the journal *Annalen der Physik* in 1817, concluding with the warning: *I consider it my duty to attract attention to the terrible effects of this new substance in order that calamity may be averted.* Clearly this advice did not reach the great majority of nineteenth century society, but the article was spotted by the great French chemist Joseph Louis Gay Lussac, who had it reprinted in French in one of the major chemical journals – the *Annales de Chimie* – and also took the liberty of renaming the substance 'morphine'. The commercial production of pure morphine was not possible until William Gregory of the University of Edinburgh perfected a method of extraction and purification in 1831, and it was then produced in bulk by Macfarlan and Co. of Edinburgh.

While the literati and others continued their love affair with laudanum, the medical profession turned increasingly to pure morphine. Administration of the drug was facilitated by the invention of the hypodermic syringe and needle by Alexander Wood of Edinburgh in 1853, and later modifications made by Charles Hunter of St George's Hospital, London, ensured that casualties of the two great wars of the mid nineteenth century – the American Civil War and the Crimean War – could receive pain relief in an efficient manner. Huge quantities of opium were also dispensed to treat dysentery, and many of the soldiers returned home with an opium habit. For many of these returning soldiers, consumption of opium was also a way of forgetting the horrors and savagery of the conflicts in which they had fought.

Both the British and American governments began to seek ways of curbing the illicit trade in opium. After several failed attempts, the new Liberal Government of 1893 tabled a motion that:

The Indo-Chinese opium trade is morally indefensible, and requests Her Majesty's Government to take such steps as might be necessary for bringing about its speedy close.

In the same year, Gladstone's government set up a Royal Commission to establish the facts about opium use and abuse, especially with regard to the production of Indian opium. The Commission heard evidence from 723 people and reported its findings in 1895. Overall it concluded that it was unnecessary to prohibit the production of Indian opium, not least because there would be a huge loss of revenue for the British Government but also because it would represent an unacceptable act of prohibition for the Indian people. The many Indian consumers of opium were reported to be:

Hale and hearty, fit for work and whose habit did not interfere with their longevity or health.

And the report further declared that *there was no such thing as a murderous opium maniac.*

The British settlements in Asia had for years been heavily dependent upon the opium trade, and despite legislation from the House of Commons in 1908, which effectively closed the opium dens in Hong Kong and Ceylon, opium use was not declared illegal until 1945 – 103 years after Hong Kong had been ceded to the British following the end of the first Opium War in 1842. Making a drug illegal does not guarantee its elimination, and in the mid 1970s there were an estimated 30,000 opium smokers and 120,000 heroin users in Hong Kong who required in excess of 35 tons of opium to satisfy their habits.

Similar grand plans for the complete eradication of opium use in China were initiated following an imperial decree in 1906 banning opium smoking and seeking the closure of all opium dens by 1917. However, these remained as merely fine intentions all the time cultivation of opium continued apace in India and in what became known as the Golden Triangle – a temperate region 1000 m above sea level and of 225,000 square kilometres between the borders of Thailand, Burma and Laos. In fact near anarchy in the opium trade existed throughout the first half of the twentieth century. This was especially true in China where local warlords controlled the trade and, following the invasion of Manchuria by the Japanese in 1931, production was further encouraged. Throughout this Japanese incursion and right through to the arrival of the communist government of Mao Tse-tung in 1950, the nationalist government of Chiang Kai-shek relied upon opium revenues for its very survival. Not surprisingly, by the end of the 1930s, possibly as much as 10% of the Chinese population were opium addicts.

Elsewhere in Southeast Asia, the French colonial forces had also been involved in the opium trade for many decades, initially in Cambodia and Laos, then later in Vietnam. An undoubted early champion of this opium trade was the first Governor-General of French Indochina, Paul Doumer, who after his appointment in 1897 set about the total reorganisation of opium production starting with the construction of a new opium refinery in Saigon. The revenues from this trade very rapidly generated enough money to pay for much of the development of their colonial territories. Interestingly, Paul Doumer entered politics on his return to France in 1902, rising to become President of France in May 1931, only to be assassinated by an unhinged Russian émigré one year later.

By the end of the Second World War production in Laos with French co-operation probably exceeded 30 tons per year, and following removal of official French Government support in 1946, French military intelligence took over support of the opium trade. These intelligence operatives developed a hugely successful liaison with the Hmong hill tribes of Laos, and these not only provided the raw opium (their only cash crop), but with French munitions acted as an effective guerrilla force (*maquis*) to combat the emerging liberation army of Vietnam. This trade took on an international complexion with the help of Corsican gangsters working out of Laos and Vietnam and collaborating with their compatriots based initially in Corsica but later in the city of Marseille. This opium was soon finding its way to the streets of New York and into some European capitals. However, 80 years of French colonial involvement in Indochina came to an abrupt end in May 1954 when a powerful Viet-Minh army of 50,000 surrounded and defeated a French army of more than 15,000 at Dien Bien Phu on the border between Laos and North Vietnam. The armistice signed in Geneva allowed the French to retain hold of South Vietnam (south of the seventeenth parallel) while an election was called to decide who would control a reunified Vietnam. However, a bloody civil war erupted in Saigon in early 1955, and the French Government finally lost its enthusiasm for its Indochina colonies and withdrew in April 1956. They left behind several hundred (mainly) Corsican gangsters who ran what became known as Air Opium, a collective name for a number of small airlines servicing the opium trade and making regular flights between Laos and South Vietnam.

Various international meetings were convened between the World Wars to try to curb the spread of narcotics, but these were mostly inconclusive or had no real power. The International Opium Conference was held in Geneva from November 1924 to January 1925 and although 36 governments were represented they only managed to agree some partial control for opium, with no measures agreed for morphine or heroin. A second Geneva conference on the limitation of the manufacture of narcotic drugs was signed by 57 nations but required a second conference called under the auspices of the League of Nations to try to provide some rules of enforcement. During this inter-war period, opium production increased dramatically in Persia, in much of the Balkans and in Turkey. Finally, the UN Fund for Drug Control was set up in 1971, but has unfortunately been as ineffective in controlling drugs as its parent body has been in its efforts to stop wars and prevent famine. These worthy bodies were, however, not operating on a level playing field because many governments, especially in Southeast Asia, relied on the revenues from the opium trade, and received assistance from various clandestine organisations which allegedly included French military intelligence and the CIA.

The arrival of the communists in China heralded the dawn of CIA involvement in Southeast Asia, because although the Americans had supported the Nationalist Army of Chiang Kia-shek (the Kuomintang or KMT) in their fight against the Japanese, there was now a need to halt the anticipated spread of communism in this part of the world. They supported the residual units of the KMT on the border between Burma and China, and the CIA airline – initially Civil Air Transport (1950) but later renamed Air America (1959) – not only supplied the KMT with arms and ammunition but also allegedly helped with the transportation of opium. Initially, the KMT made brave but futile efforts to invade China, but quickly realised that this was a fruitless enterprise and settled for the highly lucrative opium trade. In the 1950s this amounted to around 80 tons per annum but by the early 1960s total production of opium amounted to 300–400 tons per annum. The KMT were not the only producers of opium since both the Thai and Burmese governments supported a lucrative trade in the drug, and (possibly with CIA support) the military rulers in Thailand had, by 1955, turned Bangkok into one of the world's chief trading centres for opium. The total opium production of the Golden Triangle at the end of the 1960s was probably 1200–1400 tons.

The scene was set for the emergence of a new and much more dangerous trade based upon the morphine derivative heroin. This drug is an even more effective means of pain relief – and also abuse – than the parent opium or morphine, and diacetyl morphine (later to be called heroin) was first prepared by the pharmacist Alder Wright working at St Mary's Hospital, Paddington, in 1874. He was seeking a non-addictive alternative to morphine, and simply boiled morphine with acetic anhydride to produce a new crystalline product. This was administered to dogs and shown to induce drowsiness and also vomiting, and although a number of reports of this new narcotic appeared in journals, little attention was paid to the discovery. It was resurrected in 1898 by the German chemist Heinrich Dreser, chief pharmacologist of the Bayer company at Elberfeld. He was so impressed by its analgesic properties that he called it heroin based upon the German word – *heroisch*.

In addition to its utility as an extremely potent painkiller, what elevates heroin to the status of a major drug of abuse is its ease of preparation using simple and cheap chemicals. Raw opium is dissolved in hot water and then treated with lime to precipitate impurities (which are filtered off) and produce calcium morphinate which is present in the filtrate. Careful acidification then causes the morphine to precipitate – this is what is termed *morphine base* and represents about 10% of the weight of the raw opium. Now the conversion into heroin can commence, and the morphine is simply heated with acetic anhydride usually in a glass vessel for a number of hours at 85 °C to effect the conversion. Water is added, then the mixture is neutralised with sodium carbonate solution, and the heroin precipitates. This is typically sold as 'brown sugar' and contains about 30–40% of heroin so is only suitable for smoking (number 3 grade). It can be further purified by recrystallisation using alcohol and ether mixtures and activated charcoal, or converted into its water-soluble hydrochloride salt. The heroin is now of 80–100% purity and is suitable for injecting (number 4 grade). These procedures are so simple and the chemicals so readily available that it is all too easy to see how it became part of the cottage industries in Burma, Afghanistan and, in particular, the backstreets of Marseille and Hong Kong.

The involvement of heroin gangs in Marseille is familiar to most people due to the success of the films *The French Connection* and *French Connection II*, but the origins of this trade are even more fascinating than these semi-factual stories. The era of prohibition in the US had encouraged the formation of criminal gangs that dealt in alcohol and narcotics, though interestingly the Mafia were originally only involved in the first of these activities since at this time certain family values were still respected and the Mafia 'families' would not become involved in narcotics or prostitution. However, they observed the vast sums of money being made by less scrupulous syndicates and soon became involved in this growing business opportunity. From the early 1930s, a new breed of Mafiosi led by ruthless gangsters like Charles "Lucky" Luciano emerged in New York City to dominate the closely connected trade in narcotics and prostitution, and this continued essentially unchecked until the Second World War, even though Luciano was imprisoned in 1936.

The Mafia had not had such an easy time in Italy because Mussolini had initiated a crackdown in the late 1920s, and by the start of the Second World War there was only limited narcotics activity in Sicily. However, Mafia fortunes revived following the liberation of Sicily from the Germans in 1943, when the US Government's Office of Strategic Studies (OSS) – the forerunner of the Central Intelligence Agency – helped to install some of the local Mafia bosses in key political positions in order to frustrate the growing influence of the local communists. In 1946, Luciano was freed from jail and deported to Sicily. He and other senior Mafiosi were then in a position to facilitate the growth of heroin factories on the island to supply the growing demands for the drug from the USA and Canada. By 1948 most of the heroin smuggled into North America was derived from crude Turkish opium that had been shipped *via* the Lebanon to Sicily or into Marseille for chemical transformation into heroin. The Corsican narcotics gangs in Marseille had a long history in purifying crude Turkish opium and, during the period 1950 to 1970, the Mafia and Corsican syndicates agreed to share the American Market with an increasing number of laboratories established in

Marseille and controlled by the powerful Corsican families. What is remarkable is that although the French police managed to close more than a dozen of these factories, others remained in operation throughout the 1960s. While some inertia was doubtless due to police corruption in Marseille, the supposition has always been that the French Government tolerated this highly damaging industry because the Corsican syndicates were useful paid participants in violent struggles against the strong local communist unions. These unions were so strong in 1950 that they were able to completely close the port of Marseille, and this effectively prevented military supplies leaving for Indochina where the French Army was involved in its increasingly desperate struggle with the Viet-Minh. The strike was eventually broken after violent clashes with the Corsican gangsters.

At the height of their activities in Marseille during the mid 1960s, it has been estimated that there were around 20–25 laboratories in the city each producing 50–150 kg of heroin per month, most of it destined for the major cities of the USA and Canada. However, two factors conspired to end this domination of the heroin market in the early 1970s. Firstly the Turkish Government was persuaded by the Nixon Administration to introduce a successful ban on opium production, and secondly the French Government – again under pressure from the US – began to apply much greater pressure on the gangs. These were quite rapidly forced out of the city and transferred their activities to Indochina. The message from Richard Nixon to the American people and to his allies abroad concerning his war on drugs and other aberrant behaviour was simple:

Do you think the Russians allow dope? Hell no . . . homosexuality, dope, immorality in general – these are the enemies of strong societies.

It is interesting to compare the activities of the heroin gangs in Marseille with their counterparts in Hong Kong. This island outpost of the British Empire had a long history of opium production, and from the 1960s became a major centre of heroin production rivalling Marseille. At this time, much of the opium from the Golden Triangle passed through Thai ports and thence by trawler to Hong Kong for processing. Most of the criminal gangs had learnt their trade in Shanghai between the wars, and were only displaced from Shanghai when it became clear that the communist forces of Mao Tse-tung were going to occupy the city in the late 1940s. Between the wars, the so-called Green Gang (*qingbang*) monopolised the opium trade in

Shanghai with as many as 100,000 gang members by 1920, and up to 40,000 chests of Persian, Turkish, Indian and Chinese opium passing through Hong Kong each year. Most of the heroin users smoked heroin rather than injecting it since they had easy access to grade 3 drug (around 40% heroin). Lumps of the drug were typically placed onto aluminium foil and heated with a flame, then the smoke was inhaled *via* a rolled-up sheet of paper. This was known as 'chasing the dragon'.

After the end of the Second World War, the Americans were especially sensitive to the need for influence in Southeast Asia since they feared communist expansionary plans. In response to this perceived threat, the Truman administration transformed the wartime OSS into the CIA with a two-fold mission in espionage and covert operations. Interestingly, the OSS had supported the newly formed Viet-Minh and its leaders Ho Chi Minh and General Vo Nguyen Giap when they fought the Japanese during the closing months of the Second World War. This support was withdrawn once it became clear that the French were going to try to regain their colonial empire by waging an anti-communist crusade against the forces of Ho Chi Minh and General Giap. However, these French aspirations were ended at the Battle of Dien Bien Phu in 1954.

The CIA operated first in Burma during the 1950s, then Laos in the 1960s, and after the French colonial forces finally left Indochina in mid 1956, the CIA had the field to themselves. From 1959 to 1970, the CIA used the aircraft of Air America to help maintain a guerrilla army comprising as many as 45,000 of the local Hmong tribespeople. The origins of Air America are very interesting since the airline started life as the private air force of the American General Claire Chennault, whose Flying Tigers supported Chiang Kai-shek and his forces (KMT) in their struggles with the Japanese during the Second World War. At the end of the war, Chennault continued to support the KMT and other forces opposed to the communist regime, and the CIA then purchased his Civil Air Transport, renaming it Air America. The Hmong tribespeople were now tasked with putting pressure on the communist Pathet Lao forces who operated near the border of Laos and North Vietnam, and were increasingly supported by the armed forces of North Vietnam. Arms, food and medical supplies were shipped to the villages of the Hmong fighters, and opium reputedly made its way back to the cities of Long Tieng and Vientiane, and thence to Saigon and onwards to the international market. The role of Air America and the agents of the CIA in opium smuggling has been a hotly debated issue over many years. Most of the books written about the history of the CIA imply with varying levels of veracity that opium was frequently carried out of Laos on Air America planes, though CIA chiefs over the years have always denied that the pilots and CIA operatives were aware of these cargoes.[i] The gung ho activities of Mel Gibson and his fellow actors in the 1990 film *Air America* are certainly not credible and caused the CIA a serious degree of distress. What is clear is that the crews of Air America planes carried out a huge number of operations in support of the French forces at Dien Bien Phu and of the Hmong tribespeople, and in search and rescue missions during the Vietnam War. The personnel risked their lives flying into short, ill-prepared airstrips in mountainous terrain, and at least 90 of them lost their lives during the period 1959 to 1975. It is claimed that it was an Air America helicopter that appeared in one of the iconic photographs that defined the Vietnam debacle showing desperate Vietnamese trying to flee from the roof of the apartment building of the CIA chief in Saigon, just before the city fell to the forces of North Vietnam on 30th April 1975.

The success of the opium trade was dealt a bitter blow when the Americans decided to bomb the Pathet Lao into submission in 1964, and the resultant instability amongst the native opium growers led to a major fall in production. Even so in 1969 it is likely that the region of the Golden Triangle still produced around 1000 tons of raw opium, and the heroin trade was about to receive a major boost as the Americans moved into Vietnam with a vengeance.

American experts had already been playing a leading role as advisers to the South Vietnamese army, filling the vacuum left by the withdrawal of the French colonial forces in 1956, with the aim of ensuring the survival of the anti-communist government of Ngo Dinh Diem. But the US Government found itself dragged ever more deeply into the quagmire that became the Vietnam War with significant numbers of troops involved from 1963 to 1973. At its height half a million US military personnel were engaged in fighting the North Vietnamese army and their allies the Viet-Minh (called the Viet Cong by the US and the government of Diem). A very large number of the US combatants ended up smoking dope of various kinds, primarily heroin and marijuana, and the quality

---

[i] In Anthony McCoy's 700-page book *The Politics of Heroin*, now in its third edition, he gives a comprehensive account of his views on CIA complicity in the global drug trade especially in Afghanistan, Southeast Asia and South America. He spent part of 1971 with the Hmong tribespeople, so his account has to be given serious consideration. Interestingly, the most recent book (also 700 pages long) about the CIA – *Legacy of Ashes* by the journalist Tim Weiner (2007) – makes no mention at all of Air America, and the index to the book has no entries for opium, heroin or cocaine. Despite extensive coverage about CIA activities in Afghanistan, there is no mention of the number one crop of that country – *Papaver somniferum*! The contrast between these two books could not be more stark.

of the heroin was amazing in both its purity and its relative cheapness. The criminal gangs in Hong Kong were not slow to exploit the arrival of tens of thousands of potential customers, and at least 21 specialised chemistry laboratories (staffed with Hong Kong expert chemists) were installed in the Golden Triangle to service the GI market, and these reputedly produced as much as 1200 tons of heroin per year. At least 15 of these supplied the highest grade heroin, which the service personnel usually smoked in admixture with tobacco and marijuana. At the height of the conflict around 15–20% of combatants were addicted in some measure to heroin, and an official report estimated that 1 in 3 had used heroin. If they survived their tour of duty, many of the GIs took the habit home with them. It is intriguing to note that the American military has always had stringent controls over the drinking of alcohol by servicemen under age 21, yet most of the conscripts were only 19, and the sheer volume and accessibility of drugs in Vietnam inevitably led these young men to use them in place of alcohol. They were assailed on all sides by street vendors and civilian workers in their barracks, all ready to supply small quantities of heroin.

The widespread use of drugs and an overwhelming sense that the USA was losing the war led to a collapse of morale by the early 1970s, leading the then US commander General Creighton Abrahams to note in 1971:

I've got white shirts all over the place – psychologists, drugs counsellors, detox specialists, rehab. people . . . is this a goddamned army or a mental hospital!

The US forces seem to have been particularly handicapped by the activities of the Vietnamese air force and navy who probably devoted as much time to drug smuggling as they did to fighting the enemy. Indeed it has been claimed that the South Vietnamese air force C-47 transport planes flew into Saigon's main airbase loaded with Laotian opium throughout the war years.

The eventual departure of the American forces from Vietnam after January 1973 left something of a vacuum for the narcotics gangs, but they quickly adapted to this change of fortune and turned their attentions to Europe and the Americas using initially Bangkok and then Hong Kong as their gateways for exporting heroin. However, the new communist governments of Southeast Asia were not always so tolerant of opium production and refinement, and an increasing amount of opium production transferred to Mexico, Afghanistan, Pakistan and even Iran. The revolution in January 1979 that led to the overthrow of the Shah of Iran produced an unexpected boost to opium production since the Ayatollah Khomeini and his followers were much more opposed to alcohol than they were to narcotics, and the Shah's ban on opium production was reversed. However, it was Afghanistan where most of the opium was produced.

Afghanistan had been a major producer of opium for hundreds of years, but now took centre stage in both geopolitics and the drugs trade. This country of 250,000 square kilometres (about the size of Texas) has been fought over for more than 2500 years and for its many tribes and ethnic groups opium growing and the repulsion of invaders have always been a way of life. The old Soviet Union had long recognised that this unstable neighbour was a breeding ground for Muslim extremists who could easily upset the delicate balance of its own multi-ethnic state. So from 1965 the Soviet government actively supported the emerging neo-communist People's Democratic Party of Afghanistan (PDPA), and this body finally seized power following a coup in April 1978. The PDPA's proposed reforms included equal rights and education for women and major land reforms that would have overturned centuries of male-dominated socioeconomic structure. Not surprisingly a revolution began almost at once, and there were mass desertions from the Afghan Army. The Soviet leadership was now under pressure on two fronts – its favoured government was in serious trouble and they were also apprehensive about US plans for Afghanistan now that Iran had become militantly anti-American. Invasion became inevitable and the Russian army rolled into Kabul on Christmas Eve 1979, little realising that this would precipitate a holy war (*jihad*) and the mobilisation of all of the Afghan tribes and ethnic groups. From every mosque and meeting place, thousands of mullahs promulgated the eviction of the invader by whatever means. For the freedom fighters or *mujahidin* this was merely the latest and most powerful invader in their long history of conflict. As in all such conflicts powerful and charismatic leaders emerged, and the two most prominent were Ahmad Massoud, leader of what became known as the Northern Alliance, and Osama bin Laden, who led a group based in Pakistan from about 1982.

The Soviet invasion of Afghanistan was viewed by the US Government as a Soviet threat to the Persian Gulf oil fields, so the CIA and other military units were soon involved. At the height of the war as many as 30,000 mujahidin, armed with American weapons, were in combat. The CIA played the same kind of role they had in Burma, supporting the local mujahidin guerrillas with armaments as they divided their time between the opium trade and fighting the forces of the Soviet Union. As part of the hundreds of millions of dollars worth of munitions the US supplied, more than 2000 Stinger anti-aircraft missiles were given to the mujahidin. These shoulder-held, heat-seeking missiles turned the tide in the conflict, and morale in the Soviet army and air force collapsed as more than 270 aircraft were brought down in 1987 alone. Altogether, at least 14,000 Russians lost their lives during the conflict, and the real figure may be twice that number. By May 1988, the Soviet government had lost heart and a slow withdrawal began and was complete by February 1989. This left behind a power vacuum and this led inevitably to a vicious and protracted civil war. Initially, Massoud's Northern Alliance were the dominant force, but they gradually gave way to bin Laden's new Taliban, which emerged from the southern part of Afghanistan and was strongly supported by Pakistan. The Taliban finally took Kabul in September 1996, and the Northern Alliance retreated to their strongholds and poppy fields in the north of the country.

Once in command, the Taliban proceeded to impose a tax of 5–10% on all opium harvested and a separate tax on heroin production and shipment. At the same time they deplored the consumption of drugs by the Afghan population and encouraged opium and heroin exports, and the annual income for the Taliban has been estimated to have been $8 billion as Afghanistan became one of the world's leading suppliers. Periodically the Taliban leadership made overtures to the United Nations promising to eradicate opium production in return for recognition of the regime and humanitarian aid. Initially these overtures went unanswered but, following a disastrous opium harvest in 2000, Mullah Omar increased his bid for recognition by putting into force a total ban on opium production, and this was rewarded with $43 million of humanitarian aid. The subsequent ban had the effect of driving opium production into the lands of the Northern Alliance, who were still waging war against the Taliban and had always financed their military activities through sales of opium and heroin.

Interestingly, a further source of income for both the Taliban and Northern Alliance was a 'buy-back' policy for Stinger missiles introduced by the CIA following the departure of the Soviet forces. A price of $80,000 to $150,000 was offered for each recovered missile, and this policy seems to have been partially successful, though in 1996 the CIA admitted that 600 missiles were still missing. This has had the effect of making subsequent military operations in Afghanistan fraught with danger since a majority of the missing missiles are undoubtedly in the hands of the Taliban.

Following the events of 11th September 2001, and the invasion and overthrow of the Taliban forces, the Northern Alliance have been able to concentrate on opium production and there has been a resurgence in the opium trade. In 2005 it was estimated that 820 metric tons of illicit opium was produced in Afghanistan, which is more than twice the global need for medical opiates (perhaps 380 metric tons), and Afghanistan now probably supplies around 90% of the world's heroin (see Figure 2.5). Attempts by NATO forces to eradicate the opium crop have usually been counter-productive since the local farmers see this as an evil act by yet another invader. Thus far, programmes to encourage the farmers to grow alternative cash crops, like *Artemisia annua*, the source of the anti-malarial drug artemisinin, have been unsuccessful.

All of this trafficking in opium, morphine and heroin requires smuggling routes and methods of concealment, and over the years some highly imaginative schemes have been devised. Almost every imaginable hollow object has been used at one time or another including dolls, boot heels, spare tyres, bicycle frames, plaster casts and screw-in horns of cattle. The use of human 'mules' who have swallowed condoms and other sealed receptacles full of drugs is well documented, not least because of the numerous reported deaths following leakage of these containers. Insertion of such packages into the anus or vagina is also well known, though the diplomatic bag has also been used as a safer means of smuggling. The most frequently cited example of this practice involved the new Laotian ambassador to France, Prince Sopsaisana,

Figure 2.5   90% of the world's heroin originates in the poppy fields of Afghanistan.
© Associated Press.

who arrived in Paris in April 1971 to take up his post. One suitcase from his vast array of luggage was temporarily lost at Orly Airport, and when it reappeared it was opened by customs officers to reveal 60 kg of grade 4 heroin. Although this diplomatic *faux pas* was hushed up, the French refused to accept the Prince as an ambassador, and he was ultimately recalled to Laos. However, despite these individual efforts, the majority of large-scale smuggling is carried out by plane, ship or long-distance lorry (*e.g.* from Northern Afghanistan into Tajikistan, Uzbekistan, Kazakhstan and onwards through Russia and into Europe), and most countries would admit that they believe up to 90% of drugs that are shipped actually arrive safely. As for the huge sums of money generated through the sale of heroin, these are usually laundered *via* casinos, bureaux de change and other larger-scale illicit banking operations.

So the popularity of the opiates, especially heroin, is undiminished, but all of this begs the question: why are humans affected by morphine and heroin? This question remained unanswered until 1975 when John Hughes and his colleagues at the University of Aberdeen isolated two molecules from pig brain that had opiate-like activity. In fact they were up to 20 times more potent than morphine in several assays and appeared to act at the same biological sites as morphine and related opiates. Hughes went on to elucidate the chemical structures of these molecules and showed them to be pentapeptides which he christened met-enkephalin and leu-enkephalin after the names of the terminal amino acids (methionine and leucine respectively). The discovery of these two endogenous opiates was merely the tip of the iceberg, and they were soon joined by a range of other more complex molecules which fell into two broad families termed endorphins and dynorphins, all of which were polypeptides, and more recently the endomorphins which are tet-rapeptides. Thirty years later we still do not know the precise role of these mammalian opiates, although it is clear that they do act as endogenous painkillers, and probably play a role as mediators of that sense of well-being associated with exercise or the consumption of certain foods like chocolate.

All of the plant-derived opiates and these endogenous polypeptides bind to opiate receptors in the brain, gastrointestinal tract and elsewhere. The most important of these are the so-called µ-receptors, which are large proteins on the surface of responsive cells. Attachment of morphine and other opiates to these receptors results in the opening of an ion channel through the nerve cell membrane, and this allows an outpouring of potassium ions, which induces inhibition of nerve cell function. The ultimate result is pain prevention. In the gastrointestinal tract, binding of opiates to nerve cells has the effect of slowing peristalsis and leading to constipation. The mechanisms by which opiates induce euphoria and a feeling of tranquillity are not known, though there is some evidence that in certain parts of the brain they enhance the release of the neurotransmitter dopamine which is intimately involved with the pleasure responses.

Over the years there has been intense interest in the chemical synthesis of structural analogues of both morphine and the endogenous opiates. The structure of morphine itself was not established until 1925 by Robert Robinson and J. Masson Gulland, but the easy availability of (in particular) thebaine, whose highly functionalised structure lends itself best of all to chemical modification, meant that the pharmaceutical companies found it relatively easy to prepare structural analogues. Drugs like etorphine (1,000 times more potent than morphine and so potent it is used to immobilise elephants and polar bears) and buprenorphine (25–50 times more potent than morphine as an analgesic but with less potential to cause addiction) are but two examples of very useful analgesics that arose from these studies. Both of these owe their enhanced potency to their lipophilic nature and resultant ease of crossing the blood-brain barrier, and also their greater affinity for the μ-receptors. Of growing importance were the efforts to produce drugs that would antagonise the effects of morphine and these would be expected to have utility in the treatment of overdose and addiction. Naloxone and naltrexone are two highly successful examples of this type of drug, both of which attach themselves to sites in the brain and elsewhere where morphine and heroin bind, and either displace the drug (in overdose) or substitute for the drug without giving any reward.

All of these analogues have more complex structures than morphine, and there was also a systematic investigation of the reduction of this complexity, and over the years this has led to the discovery of drugs like the benzomorphan bremazocine, which has 200 times greater analgesic activity than morphine, but essentially no addictive potential. Even greater simplification led to meperidine (pethidine) with 20% of the activity of morphine. In fact pethidine has a particularly interesting history since it was first synthesised in 1930 by Otto Eisleb at Hoechst in Germany as an analogue of atropine for use as an anti-spasmodic. It was not shown to have opiate-like activity until 1939, and was then marketed as a non-addictive alternative to morphine for treatment of moderate pain, the assumption being made that lack of structural similarity to morphine would lessen the chances of addiction. In the event, pethidine produces a similar spectrum of undesirable side-effects to morphine, but its rapid onset of activity and short duration of action make it an especially useful drug for treatment of pain during childbirth.

The structurally similar methadone is the mainstay of drug rehabilitation. It has similar activity to morphine but with less severe side-effects, though it is addictive so it can only work when the drug addict has a serious desire to be weaned off opiates rather than swapping one dependency for another. Finally, one drug that will be familiar to all those who travel in tropical countries, or who frequent kebab vans, is loperamide or Imodium. This analogue of methadone acts at opiate receptors in the gut and acts as a very effective anti-diarrhoeal agent – it is 40 to 50 times more effective than morphine and has low penetration of the central nervous system.

So we are pretty well equipped with our armoury of synthetic opiates to deal with pain, addiction and overdose, but the abuse of morphine and in particular heroin goes on without any sign of diminishing. The catalogue of musicians who have paid the price of addiction or overdose is very long. The early jazz musicians of the 1930s were notorious for their heroin habits and famous addicts included Charlie (Bird) Parker, Chet Baker, Art Pepper and Stan Getz, with Billie Holliday the most famous of the female addicts. Amongst the pop stars who have been addicted are Ginger Baker and Eric Clapton of Cream, Boy George, Keith Richard of The Rolling Stones, Pete Townsend of The Who and Marianne Faithfull, while heroin contributed to the deaths of Janis Joplin, Jim Morrison of The Doors, Jerry Garcia of The Grateful Dead, Sid Vicious of The Sex Pistols, Keith Moon of The Who, Jimi Hendrix and many lesser stars. The association of drug use with the music fraternity will last as long as songs like Bob Dylan's *Rainy Day Women* with its chorus *everyone must get stoned*, or Procul Harum's *Whiter Shade of Pale* (whose lyrics still defy interpretation) retain their popularity. However, these are but modern variants of the fanciful outpouring of the lesser-known Victorian poet Arthur Symons in his *The Opium Smoker*:

> *I am engulfed, and drown deliciously.*
> *Soft music like a perfume, and sweet light,*
> *Golden with audible odours exquisite.*
> *Swathe me with cerements for eternity.*
> *Time is no more, I pause and yet I flee,*
> *A million ages wrap me round with night.*
> *I drain a million ages of delight,*
> *I hold the future in my memory.*

CHAPTER 3

Coca and Cocaine

The origins of the *acullico* or 'chew' of coca are lost in the mists of South American history. There are illustrations on Mochican pottery from around AD 500 in what is known as the Classic Epoch of civilisation along the northern coast of Peru, and these depict people with cheek bulges that probably represent a quid of chewed coca. This practice of taking leaves of the coca plant *Erythroxylum coca* (a bushy shrub that grows to a height of 1 to 2 metres), chewing them into a ball and then applying lime or a similar alkali to release the cocaine is thus at least 1500 years old. The first published account of this practice by a European was probably in a letter written by Amerigo Vespucci in 1499. He had landed on the island of Santa Margarita off the northern coast of Venezuela and described how the natives: *all had their cheeks swollen out with a green shrub inside, which they were constantly chewing like beasts.*

It was the Inca Empire that elevated the use of coca to an almost divine status. The founding fathers of this empire probably settled near Cusco in what is now Peru around the twelfth century, and by the sixteenth century the empire included nearly a million square miles of country that stretched all the way from present-day Ecuador in the north to central Chile in the south. It encompassed all of what are now Ecuador, Peru, Bolivia and parts of Chile, Colombia and Argentina – an empire that rivalled in extent that of the Romans, and like the Roman Empire the Inca Empire was criss-crossed by an excellent road network. This was a totalitarian regime with no written language and without

Turn On and Tune In: Psychedelics, Narcotics and Euphoriants
By John Mann
© John Mann 2009
Published by the Royal Society of Chemistry, www.rsc.org

wheeled transport, where human sacrifice was a common occurrence, and in which *Erythroxylum coca* was a sacred plant – a living manifestation of divine beings – and as such used mainly by the nobility. Louis Lewin in his *Phantastica: Narcotic and Stimulating Drugs* cites an Indian legend that claimed the Gods had presented Man with the coca plant to *satisfy the hungry, provide the weary and fainting with new vigour, and cause the unhappy to forget their miseries.* A more recreational use was claimed by William Mortimer in his *Divine Plant of the Incas: History of Coca* (1901) in which he described an initiation ceremony for young Inca men of the noble class (the *huaraca*) in which they competed in races and other feats of athleticism while maidens offered them coca and chicha beer while crying: *come quickly youths, for we are waiting.* An offer that would have been hard to refuse even without the cocaine and alcohol!

The arrival of the conquistadores at the heart of the Inca Empire in 1532, with their horses, wheeled transport and firearms, must have been a terrifying experience. Nonetheless the Inca people held out until 1572, though they were subjected to huge loss of life from war and imported diseases (smallpox, measles and influenza), and it is estimated that the total Inca population of perhaps 16 million was reduced by 60–90% following the Spanish invasion. The Spaniards were not slow to notice the benefits of coca consumption that included appetite suppression, and alleviation of thirst, fatigue and the sensation of cold. So while coca consumption had primarily been reserved for the Inca nobility, they now used the drug to extend the working day of the Inca slaves who worked the gold, silver and tin mines on a near-starvation diet. Based on harvest records, it has been estimated that annual consumption of coca leaves during the three centuries of the Spanish domination of Peru and Bolivia was probably around six million kilogrammes. During this period vast quantities of silver and gold were exported to Spain, but the cost to the native population was severe.

So what about the present-day situation for the *coquero*? Coca chewing is still very much part of Peruvian culture with daily consumption of around 10–100 grams of leaves with up to 300 mg of cocaine extracted into the saliva. From a population of nearly 30 million in Peru, as many as 3 million chew coca, with perhaps as many as 15 million users throughout South America. Coca is a constituent of such everyday items as chewing gum and toothpaste, and is present in a tea known as *maté de coca*. The relief of fatigue especially at high altitude is clearly beneficial, though the symptoms of prolonged use include a pallid complexion, muscular weakness and apathy. This may be due to malnutrition and overgrowth of the liver, though various claims have been made for the nutritional qualities of coca leaves (50–70% of which are usually swallowed) including suggestions that *coqueros* have increased longevity. John Lloyd in his *Treatise on Coca* (1913) provided one of the first modern accounts of the effects of coca consumption:

> *After eating a simple breakfast of ground corn porridge they* (Colombian Indians in this case) *would start with their heavy packs, weighing from seventy-five to more than one hundred pounds, strapped to their backs. All day long they travelled at a rapid gait, over steep mountain spurs and across mucky swamps, at an altitude, that to us, without any load whatever, was most exhausting. On these trips these Indians neither rested anywhere, nor ate at noon, but incessantly sucked their wads of coca throughout the entire day.*

An earlier observer, the Portuguese explorer Hipolito Unanue, noted a further property of coca in his *Disertacion sobre el aspecto, cultivo, commercio, y virtudes de la famosa planta del Peru nombrada coca* (1794):

> *Certain coqueros, 80 years of age and over, and yet capable of such prowess* (with women) *as young men in the prime of life would be proud of.*

What is clear is that consumption of coca helped to maintain the new Spanish colonies. As the administrator Juan Matienzo de Peralta noted in 1561: *If there was no coca there would be no Peru.* However, the clergy of the cathedral of Cusco, while deriving considerable revenues from the sale of coca, nonetheless denounced addiction to the drug.

Knowledge of the properties of coca leaves was known in Europe from the sixteenth century, and the tales of the various explorers like the Frenchman Charles Marie de la Condamine (who was one of the first to report the native uses of rubber, quinine and curare) and the German Alexander von Humboldt (who amongst many things reported on the use of hallucinogenic snuffs) excited the imagination of European society. At one point the *Gentleman's Magazine* of 1814 contained an editorial urging Sir Humphry Davy (as one of England's top scientists) to turn his attentions to experimentation with coca so that it could be used as a substitute for food, thus allowing people to go without eating if the mood took them. The Italian doctor Paolo Mantegazza did much to enhance awareness of its properties in the nineteenth century. He was born in Monza in 1831, and after his training spent several years in the

Argentine Republic and Paraguay practising as a doctor. He observed the use of coca in Peru and in 1859 tried it himself, and wrote of these experiences in his book *Sulle Virtu Igieniche e Medicinali delle Coca e sugli Alimenti Nervosa in Generale*. He considered coca to be a miracle drug that might help change mankind.

However, it was the introduction of Vin Mariani in 1863 that first brought coca to the European consumer. Angelo Mariani was born in Corsica in 1838 and trained to be a chemist. He certainly knew of the writings of Mantegazza, and put this knowledge to good use with the invention of Vin Mariani, which was prepared by steeping coca leaves in Bordeaux wine. This was a major commercial success throughout Europe selling under the slogan *fortifies and refreshes body and brain*. The alcoholic extract of the coca leaves was also served up in the form of Paté Mariani and Pastilles Mariani, and the success of these various preparations was as much due to the vigorous advertising campaign mounted by Mariani as from its efficacy (see Figure 3.1). Mariani supplied free samples of his product to many of the famous figures of the time, and requested a small written contribution in return. Their responses included short biographies, essays, sketches and pieces of music. H. G. Wells, Jules Verne, Alexandre Dumas, August Rodin, Gabriel Fauré and Emile Zola all responded, and the contributions from these satisfied customers were collected by Mariani over a period of years and published in the form of a thirteen-volume portfolio entitled *Figures Contemporaines*. The British Library in London has a complete set of these books and the testimonials are charming. Dumas wrote that *le vin c'est le santé*; Zola claimed that Vin Mariani was *la liqueur de vie qui allait combattre la débilité humaine*; Fauré sent an eight-bar composition; Rodin drew a sketch; and Jules Verne claimed that even one bottle of the extraordinary Vin Mariani was sufficient to ensure 100 years of life. Pope Leo XIII is also alleged to have never been without a hip flask containing Vin Mariani, and awarded Mariani the Vatican gold medal for his product.

These relatively crude extracts of coca were soon superseded by something purer and more potent. In 1853, the German pharmacist Heinrich Wackenroder had isolated a crude greenish extract from coca leaves using a mixture of alcohol and water; but it was another pharmacist, Friedrich Gaedecke, in 1855 who first isolated white crystals after evaporation of an aqueous extract of the leaves. He christened this erythroxyline after the name of the plant. Five years later, Albert Niemann of Göttingen, using Wackenroder's method of extraction, managed to obtain his own crystals, which he christened cocaine. He managed to carry out preliminary structural investigations and also

Figure 3.1   Vin Mariani poster. © Historical Picture Archive/Corbis.

studied some simple chemical manipulations of the aqueous solution of his crystals. He reported:

> *Its solutions have an alkaline reaction, a bitter taste, promote the flow of saliva and leave a peculiar numbness, followed by a sense of cold when applied to the tongue.*

It was this report that probably attracted Sigmund Freud to begin his studies with cocaine, though he was also an admirer of the works of Mantegazza. His subsequent enthusiasm was revealed in his review article about contemporary research on cocaine entitled *Über Coca* published in 1884:

> *A few minutes after taking cocaine, one experiences a certain exhilaration and feeling of lightness. One feels a certain furriness on the lips and palate, followed by a feeling of warmth in the same areas.*

It seems he was also inspired to use cocaine for amorous purposes. In a letter to his fiancée Martha Bernays he wrote:

> *Woe to you my princess when I come. I will kiss you red and feed you until you are plump. And if you are forward, you shall see who is the stronger, a gentle little girl who doesn't eat enough or a big wild man who has cocaine in his body.*

He was also impressed by the apparent lack of apparent ill effects and by reports from the eminent toxicologist Sir Robert Christison of his great feats of mountain climbing at the age of 78 and under the influence of coca:

> *Long intensive physical work is performed without any fatigue . . . this result is enjoyed without any of the unpleasant after-effects that follow exhilaration brought about by alcohol . . . absolutely no craving for further use of cocaine appears after use.*

Freud's unbridled enthusiasm led him to recommend the injection of cocaine as a cure for morphine addiction, which was of course a disaster. He was not alone in giving this dangerous advice since there had been a number of reports from the US that also suggested the value of cocaine in morphine addiction. Freud's well-documented attempts to help his friend Ernst von Fleischl-Marxow overcome his morphine addiction, which had been acquired following use of the drug for treatment of severe pain associated with an amputation, led to the poor man becoming dependent upon both drugs.

On a more positive note, Freud listed a number of other therapeutic uses of cocaine, and these included treatment of digestive disorders (*I have learned to spare myself stomach troubles by adding a small amount of cocaine to salicylate of soda*), and use of the drug for asthma, weight loss due to mercury treatment and as an aphrodisiac. Of more lasting value were Freud's instructions to his assistant Carl Koller, a house surgeon at the Vienna General Hospital, to try to exploit the pain-killing properties of cocaine. In the event, Koller demonstrated in 1884 the potent anaesthetic properties of the drug. He and a colleague experimented first with frogs, then with a rabbit and a dog, touching the corneas of their eyes with the blunt end of a needle after instillation of a cocaine solution, and observed no apparent sensation of pain or discomfort. Emboldened by these results, they instilled a solution of cocaine under their own eyelids before touching their corneas with a pinhead. Koller later reported:

We could make a dent in the cornea without the slightest awareness of touch, let alone any unpleasant sensation or reaction. With that, the discovery of local anaesthesia was completed.

The potential of the drug, not only in ophthalmology but also in general surgery was realised immediately, and the editorial in the *British Medical Journal* of 2nd November 1884 captured the excitement of the time:

Hydrochlorate of cocaine is at the present moment attracting an amount of attention rarely accorded to any therapeutic agent not of the very first rank. It may be fairly said that the news of the introduction of a new local anaesthetic has been hailed with universal satisfaction.

This led to a flurry of interest and activity in the pharmaceutical industry, and the production of cocaine by the German company Merck increased from less than one kilogramme in 1883 to more than 8000 kilogrammes in 1885. Supplies for Great Britain were obtained from coca plants that had been introduced into Ceylon in the 1880s, and there were also Dutch plantations and factories in Indonesia, especially in Java.

Across the Atlantic in Georgia, coca was about to become a household name for another reason. John Stith Pemberton was born in

Knoxville, Georgia, in 1831, and after early schooling trained to be a pharmacist. By the outbreak of the Civil War in 1861 he had established a thriving wholesale business specialising in natural remedies, but left this to serve with distinction in the Confederate Army, and was seriously wounded in a battle near Columbus, Georgia, in 1865. This wartime trauma and probable addiction to opium was followed by post-war bankruptcy that left him determined to make his mark in business, and this he did with huge success when he introduced Pemberton's French Wine Coca in the early 1880s. He freely admitted that this was based on Vin Mariani but claimed that his product was superior in its effects as an *ideal nerve tonic, health restorer and stimulant* – or so the advertisements proclaimed. Initially his drink contained coca leaf extract, kola nut extract and wine, but following the imposition of prohibition in Georgia in 1886, the wine was replaced with sugar syrup. This new beverage was he claimed:

A direct tonic to the nerves and muscles, and acts so happily on the brain that laborious thinkers and students who have used it have given it the name of 'intellectual beverage'.

In a more fanciful claim he tried to enhance its appeal even further:

Coca is a most wonderful invigorator of the sexual organs and will cure seminal weakness, impotence etc. when all other remedies fail.

This was heady stuff in America's bible belt, and searching around for a more commercial name he called his beverage *Coca-Cola* and made his fortune. The name, however, was almost lost in 1904 when the US Government became alarmed about the addictive properties of cocaine and forced the manufacturers to remove the extracts of coca. They also tried to make the Coca-Cola Company change the name but after extensive legal wrangling the company won the day, the courts ruling that the names Coca-Cola and 'coke' had become part of the American language and to ban them would be seen as inappropriate use of government powers.

The first cocaine 'epidemic' in the US probably started around 1880 and ran until the 1930s, and for much of this time cocaine was readily available from pharmacists and also found its way into a plethora of patent medicines with the cocaine content ranging from around 1% to a staggering 35% in Agnew's Catarrh Powder. There were even 'cocaine parlors', and as the abuse of the drug grew in scale, the need for legislation became urgent. Initially the Pure Food and Drug

Act of 1906 legislated for the removal of cocaine (and other supposed poisons) from patent medicines, then the passing of the Harrison Narcotic Act in 1914 identified the dangerous nature of the drug. However, many US states had by then introduced their own legislation curtailing the sale of a drug that became increasingly associated with crime, especially criminal acts carried out by what the newspapers described as 'cocaine-crazed negroes'. In the UK, the Dangerous Drugs Act of 1920 eliminated easy access to cocaine and other drugs, though they had earlier (in 1916) been denied to serving members of the armed forces as part of the Defence of the Realm Act. The drug was still readily available (at a price) in London and the other major cities, not least because there was a huge number of unlicensed dentists who had a legitimate need for cocaine. Around the world most countries introduced controlling legislation and by the 1930s the first 'epidemic' of cocaine use was over, and increasingly drug users turned to cheaper alternatives like the amphetamines. The second wave of cocaine abuse began in association with the new freedoms and evanescent pop culture of the 1960s.

For much of the twentieth century cocaine hydrochloride was administered by snorting or sniffing (scientific term – insufflation) producing an immediate 'rush' and an optimum effect in about 10–15 minutes. The cocaine salts were easy to produce and involved maceration of the coca leaves in a dilute solution of hydrochloric acid, followed by separation of the leaf mass and evaporation of the water to produce an off-white residue. In this way from 100 kg of leaves 1 kg of cocaine hydrochloride of purity 40–65% could be obtained.

The insufflation of 'lines of coke' is most typically effected using a rolled bank note, a straw, a pen casing or even a tampon applicator. However, regular use of cocaine salts by this route of administration inevitably leads to rhinitis and eventual destruction of the nasal septum. In the early 1970s, so-called 'free-basing' started to replace cocaine hydrochloride snorting. The free base was produced by neutralising the acid moiety (sulphate or hydrochloride) using a weak base like ammonia or sodium bicarbonate to provide an off-white precipitate of cocaine that could be separated by filtration or extracted into diethyl ether. This was then melted and the fumes were inhaled *via* a water pipe leading to a more intense and rapid neural response, since the large surface area of the lungs ensured an efficient access to the bloodstream and then the brain. Soon after, in 1983, 'crack' cocaine became freely available, and this was prepared by simply mixing cocaine salts and sodium bicarbonate in water then evaporating the water to produce a rock-like mass of cocaine which was then smoked. The crackling sound emanating from the hot residue inspired the name 'crack' cocaine. Smoking crack cocaine induces an effect within 4–6 seconds described as the 'cocaine rush', and this is usually associated with euphoria and pleasure compared to the best sexual experiences. Both these new types of cocaine led to rapid dependence on the drug.

Such dependence to cocaine obviously required a plentiful supply of the drug and the drug barons of South America were very happy to oblige. It has been estimated that the annual coca harvest (the new leaves can be collected as often as four times per year) in Peru and Bolivia between 1970 and the late 1980s rose from 30,000 tons to nearly 400,000 tons. Coca leaves are usually dried in the sun then chopped (often with a strimmer), before being placed in a large drum and covered with kerosene (or other solvent). After several hours of extraction, the kerosene is siphoned off and the cocaine is extracted into dilute sulphuric acid (as its salt). Relatively pure cocaine can then be precipitated by neutralisation. The resultant paste or *pasta* is then generally passed on to Colombia for processing into pure cocaine or its hydrochloride salt. In the 1970s and 1980s this was then sent to the Caribbean or the Bahamas for onwards shipment by fast boat or light plane to Florida, though these days it is often exported to West Africa for onwards transmission. By the early 1980s as many as 80 planes were making drugs drops over Florida every night, fuelling the new yuppie cocaine culture that was sweeping most US cities. At this time around 75% of the cocaine used in the USA originated in Colombia. A manifestation of the enthusiasm for this new 'cocaine epidemic' was the appearance of a host of expensive insufflation aids like gold-plated straws and razor blades, which became the new 'must-have' accessories. In 1981, President Ronald Reagan and Vice-President George Bush (senior) officially launched their War on Drugs directed at the South American cartels. Initially the US targeted the Florida route of importation, so the cartels switched to a new route *via* Mexico, and by 1994 the revenues being earned by the criminal gangs in Mexico exceeded $30 billion which dwarfed the country's oil exports of not quite $10 billion.

During the 1970s and 1980s, one man dominated the trafficking of drugs from Colombia – Pablo Escobar, head of the Medellín cartel. He was born in a suburb of Medellín in December 1949 and, after an early career in petty street crime (stealing cars and protection rackets), he graduated to become one of the most murderous and successful drug dealers in South American history. An early brush with law enforcement agents in 1976 provided a cogent signal of how he proposed to operate. He was arrested in possession of about 40 kg of cocaine, but this did not lead to a conviction because the arresting officers and the judge who sent him for trial were all murdered soon after his arrest. His operations were so successful by the mid 1980s that it is claimed he was making $1 million per day from his activities, and the whole South American cocaine business was reputed to have had a larger annual income than giant American corporations like Boeing and Proctor & Gamble. Not surprisingly he allowed himself a lavish lifestyle at his luxury ranch, which had its own airstrip, but he also liked to think of himself as a South American Robin Hood and built homes for the poor and provided them with food, medical supplies and sports facilities. He was absolutely ruthless with a policy that was usually described as *plata o plomo*, that is silver (for bribes) or lead (bullets) where bribes were not accepted.

Campaigning anti-drugs politicians and police chiefs did not usually survive for very long, though the Justice Minister Rodrigo Lara Bonilla was particularly assiduous in his pursuit of Escobar. In March 1984 he masterminded raids on several of Escobar's processing operations, seizing more than 8000 kg of high-purity cocaine worth about $1 billion in the USA. However, even Bonilla's luck eventually ran out and one of Escobar's assassination squads caught up with him in rush-hour traffic in Bogotá later that month and ended his brief career. This assassination of a popular politician did, however, precipitate an outpouring of popular revulsion in both Colombia and the USA, and the Colombian Government was forced into wide-ranging anti-drugs activities. Unfortunately this simply incensed Escobar and a near civil war situation existed for much of the rest of the decade with Escobar's henchmen bombing and murdering politicians, judges and the police with apparent impunity. They also fought a running battle with the increasingly powerful Cali cartel, who were their main Colombian competitors. On 25th March 1989, Escobar's henchmen were responsible for a synchronised bombing campaign in Cali, Medellín and Bogotá aimed at policemen, politicians and rival cartel operatives. Inevitably there were numerous civilian casualties as well.

By September 1990 the Government (and the people of Colombia) had suffered enough and an amnesty was declared. The deal stipulated that if Escobar and the other drug cartel members gave themselves up, they would receive lenient treatment and most importantly would not face deportation to the USA where the CIA was desperate to get its hands on them. In the event, Escobar surrendered in June 1991 and was then incarcerated along with his henchmen in a luxurious prison (appropriately called *la Catedral*) of his own design, which included a gym and chalet accommodation for visiting (female) visitors. He was even allowed out to watch local football matches and for shopping expeditions. In prison he carried on business as usual including organisation of the cocaine trade and contract killing. Eventually the Government could take no more and they sought to move him to one of their own less salubrious gaols, but Escobar fled from his own prison in July 1992 and went on the run in and around Medellín. The Cali cartel now helpfully set up a group of vigilantes – *los PEPES* (*Perseguidos por Pablo Escobar*) – to hunt him down with a tempting reward of $5 million attached. Escobar's friends and associates, especially his many lawyers, were rapidly eliminated. From November 1993 the Government began to use Escobar's wife and children as the bait for their trap, and the net began to close around Escobar himself. The events surrounding his eventual capture and death are still the subject of folklore, but probably involved a team that included Special Forces from both the US and Colombian armies with sophisticated communications equipment. What is clear is that he was located because he was careless with his use of a radiophone, and he died in a shoot-out (or was assassinated or even committed suicide – he received at least one bullet to the head) on 2nd December 1993, and the Medellín cartel promptly folded in his absence. But, as with most drug operations, there were others waiting to step into the breech and there is little sign in 2009 that the flow of cocaine from Colombia has been halted.

The Cali cartel certainly became a major new force not only for cocaine but increasingly for the growing Colombian opium industry. The cartel had always been a little more subtle in its methods than the Medellín cartel. They bought their enemies rather than killing them, and at one point in the mid 1990s they allegedly controlled one-third of the Colombian Congress. Their distribution networks also had greater sophistication. Small groups (cells) of distributors operated in most big US cities, and these avoided detection by using a large number of mobile phones which were discarded after just a few calls. A code of silence was also enforced through a campaign of threats against the families of the traffickers. By the turn of the century, the Colombian Government claimed to have imprisoned most of the Cali cartel members, though as much as 80% of the world's cocaine still originates in Colombia.

Another 'star' of the cocaine trade was Manuel Noriega, who was born in Panama in 1934 and, although brought up under poor circumstances by foster parents, subsequently received most of his education in military academies in Peru and the USA. By 1981 he was head of the Panamanian armed forces and effectively controlling the country, presidents notwithstanding. He was also undoubtedly supported at this time by the CIA. This power base allowed him essentially complete control over the export of Colombian cocaine across the Panamanian border and a domination of money-laundering services for the drug money flowing back across the border. He also sanctioned the construction of a huge cocaine processing factory for the Medellín cartel in the Darien jungle just inside Panama. The CIA had been active in Colombia since the 1960s, and was initially involved in the training of the police in counter-terrorist skills. Later they were involved in supporting the Contras in Honduras in their attempts to destabilise the Sandinista regime in neighbouring Nicaragua and Costa Rica. It is not clear whether, as in Indochina, their activities extended to supporting the vast drug trafficking operations of the Contras. Nonetheless a large quantity of American arms and equipment passed to the Contras *via* Panama with Noriega's help. However, when the extent of Noriega's involvement in the cocaine trade became apparent, a US invasion force of 24,000 service personnel was sent to Panama City in December 1989 to arrest him. The story now enters the realms of farce since Noriega sought refuge within the Papal Nuncio and was only prised from it after the American forces bombarded the building with non-stop, very loud pop music for several days. This military adventure cost the lives of 23 American servicemen and hundreds of innocent residents of Panama City. Noriega was subsequently tried in the US and is still serving a long prison sentence with the prospect of extradition to France for a further trial upon his release.

The funding for the various US operations in Central America was initially approved by Congress as part of the War on Drugs, but this august body later got cold feet when they better appreciated that US-backed forces were trying to overthrow a legally elected government in Nicaragua. In order to make good the shortfall in funding, the CIA appointed Colonel Oliver North, who had served with distinction in the Marine Corps in Vietnam, to take charge of fund-raising for the Contras. This he did by selling TOW anti-tank missiles to Iran and ploughing back the profits into the Contra activities. The shooting down (by the Sandinistas) in October 1986 of a C-123 transport plane loaded with arms and incriminating documents, not to mention three American crewmen, opened the floodgates of revelations about the Iran–Contra connection and led eventually to Senator John Kerry's so-called Iran-gate investigation. Almost inevitably clear evidence of Contra involvement in cocaine smuggling emerged and, although CIA involvement in these activities was never established, the whole sorry affair provides yet another example of America's unhappy experiences with the shady organisations it has supported, not to mention the countries like Iran that ended up on its blacklist (Axis of Evil). Kerry's committee concluded in its report of April 1989:

In the name of helping the Contras . . . the credibility of governmental institutions has been jeopardised by the Administration's decision to turn a blind eye to domestic and foreign corruption associated with the international narcotics' trade.

Interestingly, Oliver North was eventually cleared of all charges due to certain procedural irregularities, and he has since carved out a successful career for himself on television and as an author. His latest book, published in May 2008, was entitled *American Heroes* and contains, amongst other things, his views on fighting global terrorism!

From the early 1990s, the American military became more involved in what was termed Plan Colombia and this involved funding for military intervention and aid in the ratio of about 9:1. A programme of defoliation commenced in Peru and Bolivia. This had the effect of shifting coca production to Colombia and the area under cultivation increased from 120,000 hectares in 1999 to more than 170,000 hectares in 2001, at which point the defoliation campaign switched to Colombia. Despite these major military campaigns, the street price of cocaine has continued to fall, and the typical cost in the USA is now a mere 20% of what it was in real terms in the 1980s. It is little surprise then that there are at least 6 million users of cocaine in the USA, to add to the half a million users of heroin, and the more than 25 million users of marijuana. Certainly there is little evidence to suggest that the *Just say no* campaign launched by Nancy and Ronald Reagan in 1986 has had much effect. They famously exhorted youth to *Just say Yes to your life, and when it comes to alcohol and drugs, just say No.*

In the UK the problem has never been so serious, though the relentless campaign waged by the Thatcher government in the 1980s probably drove the criminal gangs into becoming more efficient and professional. She famously warned them *We are after you. The pursuit will be relentless. We shall make your life not worth living.* But it is certain that many of the drugs barons have outlasted her time in office, and there was never a shortage of cocaine for the City yuppies of the 1980s and 1990s.

There is a very extensive literature on cocaine use and abuse, and a good selection of modern writing on the subject of abuse and addiction is contained within the pages of *White Lines* by Stephen Hyde and Geno Zanetti. This contains extracts from the writings of amongst others Aleister Crowley, William Burroughs, Miles Davis, Carrie Fisher,

Stephen King, Will Self, J. G. Ballard and Irvine Welsh, and in the main the extracts make very depressing reading. Aleister Crowley is the exception and he extolled the aphrodisiac virtues of cocaine:

> *I had never been particularly keen on women . . . but with cocaine things are absolutely different. Until you've got your mouth full of cocaine, you don't know what kissing is. One kiss goes on from phase to phase and you never get tired . . . you're on full speed all the time* [from *Diary of a Drug Fiend*].

The excesses of Hollywood in the yuppie eighties are eloquently described by Charles Fleming:

> *Avoiding narcotics in Hollywood in the 1980s was as difficult as avoiding the Southern California sun. Drugs were everywhere, and cocaine was king, the champagne of the recreational stimulants . . . The social uses of cocaine were myriad: it was a terrific icebreaker, and instant pick-me-up and a great come-on. Everything went better with coke. It stimulated conversation and intellectual activity, and increased sexual appetite and prowess. It didn't leave telltale odours or next-day hangovers. It wasn't passé, like marijuana or martinis, and it wasn't deadly like heroin. It was portable and easy to hide. Ingesting it required no 'kit', no special equipment. And, as an added bonus, it was just expensive enough that the hoi polloi couldn't afford it.*

An excellent account of the activities of a 1970s cocaine smuggler is provided in the novel *Snowblind* by Robert Sabbag, which contains an often hilarious account of the brief career of Zachary Swan, alias for a real and very successful coke runner. The style of the writing owes much to the hippy culture of the time, but a short extract will give something of the flavour of the book, and also an idea of the excesses that were very often the norm. Here Swan describes how he and his friends use some of the cocaine that he has just received from Colombia hidden in a carved wooden tribal head:

> *The skull yielded up 500 grams of high grade uncut cocaine, double wrapped in clear plastic . . . and the coke was* (soon) *performing fabulous and outlawed miracles in the heads of Swan, his girlfriend* (and two Australian friends). *The Bolivian brain food they had ingested was only one course in a sublime international feast which featured French wine, English gin, Lebanese hashish, Colombian*

cannabis, and a popular American synthetic known pharmacologically as methaqualone . . . The party went out of control somewhere in the early hours before dawn, and the steps he had taken to mimimize his losses were eventually undermined by the immutable laws of chemistry – his mind, simply, had turned to soup.

Perhaps the most famous fictional cocaine user was Sherlock Holmes, here describing his habit in *The Sign of Four:*

I suppose that its influence is physically a bad one. I find it, however, so transcendingly stimulating and clarifying to the mind that its second action is a matter of small moment.

There is no evidence that Sir Arthur Conan Doyle ever used cocaine but interestingly, as part of his training to be a doctor, he spent one year (1890) in Vienna studying ophthalmology. It seems highly probable that he heard about the discoveries of Freud and Koller which had been made in that city just six years earlier, and that this inspired him to depict the use of the drug in his books.

A greater mystery surrounds the writing of the *Strange Case of Dr Jekyll and Mr Hyde* by Robert Louis Stevenson in 1885. The first draft of this tale of split personality induced by a drug was apparently written in a frenzied period of six days while Stevenson, who was plagued with lung disease for much of his life, was recovering from a haemorrhage. There is no evidence that he took cocaine but once again he would have read about the new excitement surrounding the drug, and wanted to describe the split personality of a cocaine addict. The book went on to sell 40,000 copies in the first six months following publication, and by 1901 had sold more than 250,000 copies.

In the 1970s and 80s, the music industry was full of people who admitted using cocaine, often to excess, and famous examples included Marvin Gaye (*blow is what really makes me fly*), Freddie Mercury, Elton John and Phil Spector – whose Christmas card once showed a picture of him snorting coke with the legend *a little snow at Christmas has never hurt anyone*. Numerous references to cocaine use appear in the lyrics of groups like The Rolling Stones and The Eagles, and the popular 1969 film *Easy Rider*, starring Peter Fonda, Dennis Hopper and Jack Nicholson, with its explicit portrayal of LSD, hashish and cocaine use, did much to establish drug use in popular culture.

So what of the useful medicinal properties of cocaine? In the absence of any firm structural information about the drug, several pharmaceutical companies designed drugs in the latter decades of the nineteenth century with structures (piperidine ring, benzoate and methyl ester moieties) that they believed (correctly as it turned out) to be present within the cocaine molecule. Somewhat surprisingly, several of these companies had good successes, and benzocaine was introduced in 1890 and is still used in throat lozenges. Of more lasting value was procaine (introduced in 1905), later given the trade name Novocaine, and this dominated the local anaesthetic market until the middle of the twentieth century. This has now been largely superseded by lidocaine (Xylocaine) (from 1948) and it is now the most widely used local anaesthetic.

As to the mode of action of cocaine, its anaesthetic effect is caused by a blocking of the so-called sodium channels in nerve axons and this eliminates nerve conduction. Its effects on the brain seems to be mainly exerted in the region known as the nuclear accumbens and this area is usually associated with pleasurable responses ('reward centre') to stimuli like food and sex. Cocaine attaches to the dopamine transporter whose job is to control the re-uptake of the neurotransmitter into the neurones that release dopamine. As a result of this blockade, dopamine accumulates and the 'reward' response is prolonged and intensified. There are similar though less significant effects on serotonin and noradrenaline re-uptake. Cocaine is reputed to intensify and extend the pleasures of sex (especially orgasm), and most habitual users claim this as one of its most significant effects. The need to take more and more cocaine is usually due to the effects of the 'crash' that occurs after the initial 'cocaine rush' (of euphoria). The plasma half-life of cocaine is only about 50 minutes, and inhaled cocaine needs replenishment after 10–30 minutes. The overwhelming need to take more coke is nicely summarised by Zachary Swan in Robert Sabbag's *Snowblind:*

> *And so, at the zenith of the cocaine ecliptic, activity was resumed. What goes up must come down; what comes down must come up. QED. Physics, man. But now, with four people behind some heavy coke, anything was possible.*

Chronic use can lead to depression and even suicidal tendencies, and this is probably due to changes in the production or sensitivity to dopamine and serotonin. However, a large number of people use cocaine in small amounts occasionally and recreationally, and many would agree with the actress Tallulah Bankhead who famously claimed:

> *Cocaine isn't habit forming. I should know, I've been using it for years.*

CHAPTER 4

Cannabis: Hashish, Marijuana, Charas and Bhang

Quand on fait la moisson du chanvre, il se passé quelquefois d'étranges
phénomènes dans la person des travailleurs . . . la tête du moisson-
neur est pleine de tourbillons, d'autres fois elle est chargée de rêverie.

This was Charles Baudelaire's description in his book *Les Paradis Artifi-ciels* of the effects of harvesting hemp (*moisson du chanvre*) on the har-vesters. The whirlwinds (*tourbillons*) and the daydreams (*rêverie*) were apparently being caused by ingestion (perhaps deliberate) of tetrahydr-ocannabinol from the flower heads of the plant – *Cannabis sativa*. Salman Rushdie described a similar intoxication in his novel *Midnight's Children*:

But now the wind changed, and began to blow from the north: a cold
wind, and also an intoxicating one, because in the north of Kif were the
best hashish fields in the land, and at this time of year the female plants
were ripe and in heat. The air was filled with the perfume of the heady
lust of the plants, and all who breathed it become doped to some extent.

Such exposure to the major psychoactive constituent of the plant was nothing new and hemp seeds have been found in copper vessels asso-ciated with Scythian burial sites dating from at least 500–300 BC.

Turn On and Tune In: Psychedelics, Narcotics and Euphoriants
By John Mann
© John Mann 2009
Published by the Royal Society of Chemistry, www.rsc.org

Herodotus, writing in *ca.* 450 BC, talks of a tent-like structure in which the Scythians would use:

A dish placed upon the ground into which they put a number of red hot stones and then add some hempseed . . . immediately it smokes and gives out such a vapour as no Grecian vapour bath can exceed; the Scythians, delighted, shout for joy.

There is a likely error here since hemp seed does not contain any psychoactive constituents, but he was probably confusing hemp seed with the flower heads or leaves, both of which contain cannabinoids – the psychoactive components of cannabis species. Three distinct species can be identified – *Cannabis sativa*, which is a bushy plant and can grow to a height of 6 metres, and the smaller sub-species *Cannabis indica* and *Cannabis ruderalis*. Recent genetic studies indicate that *Cannabis sativa* probably arose in Kazakhstan while *Cannabis indica* probably first appeared in the western Himalayas. But long before the hemp plant was used for mind-expanding activities, various ancient civilisations had discovered the value of hemp fibres from the thick stems of mature cannabis plants for manufacturing cords and ropes, and for cloth and paper making. These activities certainly go back as far as 4000 BC in China and 3000 BC in Turkestan. The seed oil was also valued as a food source, and the medicinal properties of the plant are mentioned in the Shen Nung herbal published in the first century AD, but relating to the activities of the great Chinese herbalist Shen Nung who appears to have practised around 2800 BC. The Zend-Avesta, an ancient Persian religious text from about 600 BC, actually describes an intoxicating resin from hemp, and various historians have wrongly identified this as the *soma* of ancient Indian culture. Certainly extracts from the cannabis plant known as *bhang* have been used as part of Hindu and Buddhist religious activities for hundreds of years, and as a component of Ayurvedic medicine since at least 1500 BC. The Ebers Papyrus, which also dates from around 1500 BC, probably mentions hemp (*shemshemet*) for the treatment of pain and fever.

Bhang is just one of many terms used to describe preparations from the plant and refers to ground leaves and flowers, often mixed with milk and spices and gently heated, and then consumed as a beverage or in the form of sweets. The Indians also use another form which contains resin from the trichomes (hair-like projections from various parts of the plant), with mature flowers of female plants providing the richest source of resin. The resin may be collected by scraping the trichomes and then sifting, or through the use of leather aprons worn by the harvesters as they walk through the crop of cannabis. This method of collection of the sticky exudate is similar to the collection of opium after the poppy capsules have been incised and allowed to 'bleed'. In either case, this resinous form is the most potent form of cannabis and is known in India as *charas* and more commonly elsewhere as *hashish*, and is taken with food or smoked. Finally, the ground flowering tops and leaves are consumed with food or smoked (often with tobacco), and in Africa this is usually known as *ganja* and elsewhere as *marijuana*. This has widely differing potencies depending on the source of the plants and the content of resin from the flower tops.

The use of hashish as a euphoriant was certainly known to the Romans, and Galen wrote in AD 160 of hemp cakes that produced a sense of well-being when taken in moderation, and intoxication if taken to excess. In Persia and much of the Middle East hashish was widely used as an alternative to the much-hated alcohol since its consumption with food did not lead to intoxication, which was strictly forbidden in the Koran. Its use is mentioned several times in *One Thousand and One Arabian Nights*. It was certainly employed as a medicine by the time of the famous Persian physician Rhazes (tenth century AD), and through the centuries claims have been made for its efficacy for conditions ranging from dandruff and dysentery to tuberculosis, leprosy and venereal disease. Success in the treatment of these conditions was probably illusory, though its use as a stupefacient during childbirth and for sleep induction were undoubtedly valid. Arab traders and invaders are generally believed to have exported the use of hashish into Africa, while their invasion of Spain in the eighth century ensured that the more potent strains of the plant entered Southern Europe. In Northern Europe hemp had been cultivated from about AD 400 for rope and cloth manufacture, and was used for paper making from the twelfth century. In the time of the Tudors it was valued so highly for the manufacture of rigging for naval vessels that both Henry VIII and Elizabeth I issued decrees requiring landowners to set aside a portion of their land for hemp cultivation. Fines were imposed (as much as 5 shillings during Elizabeth's reign) for failure to comply, and for this reason and also because the production process was both tedious and very malodorous, cultivation of hemp became very unpopular. By the mid-seventeenth century England was obtaining most of its supply of hemp from Russia. Nonetheless, the overwhelming might of the British navy was at least partly due to the fine rigging on the ships. In Venice, there was a 300-metre long factory for rope making (*corderia*) in the naval base – the Arsenale – and this helped to ensure that the Venetian merchant and naval fleets were the most powerful in the Mediterranean. Hemp was also introduced into North America by the European settlers and into

South America by the conquistadores, and in both instances it was used more or less exclusively for rope and cloth making.

The first stories of its use as a euphoriant were brought to Europe by soldiers returning from the Crusades, and hashish was also inextricably linked to the cult of the *Hashshashin*. These were the fanatical assassins of the eleventh and twelfth centuries initially under the leadership of the warlord Al-Hasan ibn al-Sabah or Aloidin. He ruled over large swathes of Persia but his main stronghold occupied the fortress of Alamut in the Elburz Mountains of northern Persia. The chroniclers of the times were very much in awe of this elite band of Islamic terrorists and numerous fanciful stories were created about the recruitment and maintenance of these fanatics. The fortress of Alamut was rumoured to be a wondrous place, complete with a superb garden containing four fountains that provided wine, milk, honey or water, and was also well stocked with beautiful women. In short, a kind of earthly paradise. Marco Polo travelled in this region around 1271 and reported some of the stories that he heard in his book entitled *Il Milione*, of which the most quoted edition was published in 1330. Much of his account is now believed to be fanciful and exaggerated, but he does mention the Old Man of the Mountain (*Il Veglio della Montagna*) and the drugs that were used by the *Hashshashin* (*assassini* in Italian). These induced sleep – *li fa loro dare beveraggio che dormono* – and made the Hashshashin dream about Paradise. His conclusion was that hashish had been employed, though it is more likely that opium would have caused these effects. In any event, his stories and those of the returning crusaders brought hashish to the attention of Europeans for the first time.

Widespread use of cannabis as a euphoriant in Northern Europe did not begin until the time of the Napoleonic Wars. Napoleon took his army into Egypt in June 1798 and, although his land campaign was successful, Nelson discovered Napoleon's fleet in Aboukir Bay on 1st August and all but annihilated it in what is usually known as the Battle of the Nile. This effectively marooned Napoleon and his army in North Africa, where hashish use was widespread, and with limited access to alcohol the troops developed a penchant for the various foods and drinks containing hashish. When the remnants of a demoralised French army eventually left Egypt in the summer of 1800, they took their new habit and supplies of hashish with them to France.

Experimentation with this new drug spread slowly since supplies had to be obtained from North Africa or the Middle East. In order to supply the growing market, trade routes from Turkey and the Middle East were established with shipments by camel train across the Sinai Desert or by ship to Alexandria, and this business grew enormously throughout the nineteenth and the early part of the twentieth century. The first written account of hashish use in France was provided by Dr Jacques-Joseph Moreau in 1845. In his book, *Du Haschisch et de l'Alienation Mentale*, he described his observations made during travels in Egypt and the Middle East in the period 1837 to 1840, and his subsequent experiments with mentally ill patients in the Paris area. He provided one of the most detailed descriptions of the potential for hashish to induce euphoria, distortions of time and space, enhanced oral and visual acuity, and hallucinations. Perhaps more significantly he was the initial supplier of hashish to a group of French literati who formed themselves into *Le Club des Haschischins* from 1844. The hashish was apparently in the form of a green paste (they described it as a *confiture* or *pâté*), containing hashish, butter and various ground nuts and spices. The group used a variety of psychoactive drugs and met on a regular basis in a Paris hotel – Hôtel Pimodan – to experiment with these substances and share their experiences. Accounts of these meetings were recorded by the poet Théophile Gautier in a short piece entitled *Le club des haschischins* in 1846 and also in greater detail by the poet Charles Baudelaire in his famous book *Les Paradis Artificiels*.

Théophile Gautier writing of his first experience of hashish recorded that he had been given food that was the same as used by the Assassins: *un morceau de pâté verte ... précisément la même que le Vieux de la Montagne ingérait à ses fanatiques*, which he goes on to confirm as hashish. This caused a complete transformation in his senses with water tasting like wine and meat resembling raspberries:

J'avais éprouvé une transposition complète de goût. L'eau que je buvais me semblait avoir la saveur du vin le plus exquis, la viande se changeait dans ma bouche en framboise.

And the cries of ecstasy from the other members of the Club were fanciful in the extreme:

Mon Dieu, que je suis heureux! Quelle félicité! Je nage dans l'extase! Je suis en paradis! Je plonge dans des abîmes (chasms) de délices!

Charles Baudelaire was much more expansive and described the preparation of the hashish that they consumed:

La composition du haschisch est faite d'une décoction de chanvre indien (Cannabis indica), *de beurre e d'une petite quantité d'opium – voici une confiture verte, singulièrement odorante.*

He also identified three distinct phases of response to the drug. In the first there was much laughter, hilarity and spatial distortion:

> *Les éclats de rire* (bursts of laughter), *les énormités incompréhensibles, les jeux de mots inextricables, les gestes baroques* (fantastic movements).

For him the second phase was clearly the most interesting with sounds and colours and numbers becoming mixed up, and frighteningly difficult mathematical puzzles becoming explicable. The sense of time also disappeared:

> *Les hallucinations commencent. Les objets extérieurs prennent des apparences monstreuses . . . . Les sons ont une couleur, les couleurs ont une musique. Les notes musicales sont des nombres, et vous résolvez avec une rapidité effrayantes prodigieux calculs arithmétiques.*

Finally in the third phase there was a feeling of calm and feelings of love and artistic awareness took on a unique form:

> *Dans ce suprême état, l'amour chez les esprits tendres et artistiques, prend les formes les plus singulières.*

However, at the end of this apparent passionate endorsement of hashish, he drops the bombshell that he really prefers wine!

> *Le vin exalte la volonté* (the will); *le haschisch l'annihile. Le vin est un support physique; le haschisch est une arme pour le suicide. Le vin rend bon et sociable; le haschisch est isolant* (insulating).

All of this has to be taken with a degree of scepticism since, like the pop stars of later years, Gautier and Baudelaire and the other members of Le Club des Haschichins were consuming a mixture of drugs and probably alcohol as well. Baudelaire at the very least was both an alcoholic and an opium addict, though his death in 1867 aged 46 was blamed on the syphilis that he had contracted as a young man. In any event, Le Club did not last long and by 1849 the members had dispersed. The experiences of the other members – the painter Eugène Delacroix and the writers Gustave Flaubert, Honorè de Balzac and Alexandre Dumas – were not recorded in detail. Dumas does describe the supposed aphrodisiac effects of hashish on one of his characters, Baron Franz d'Epinay, in *The Count of Monte Cristo*. The Baron, as a guest of Sinbad, alias the Count, at his hideaway on the Island of Monte Cristo, was given a green hashish paste to eat and reports the following:

> *His body seemed to acquire the lightness of some immaterial being, his mind became unimaginably clear and his senses seemed to double their faculties.*

He was then overcome with lust for some marble statues:

> *This lust was almost pain and this voluptuousness almost torture, as he felt the lips of these statues, supple and cold as the coils of a viper, touched his parched mouth.*

Elsewhere in the world other writers and poets were also experimenting. In the USA, Fitz Hugh Ludlow, a pastor's son from New York City, took increasing doses of a tincture from *Cannabis indica* for more than 15 years prior to his death in 1870 and wrote of his experiences in *The Hasheesh Eater, being Passages from the Life of a Pythagorean*. This seems to have been a blatant attempt to emulate the success of de Quincey's *Memoirs of an Opium Eater*, but is considered to be a very inferior example of the genre, though it did raise awareness in the USA of the psychoactive properties of hashish. Louisa M. Alcott, author of that archetypal book on American womanhood – *Little Women* – also wrote a short story in 1869 entitled *Perilous Play*, which famously begins with one of the female characters declaring: *If someone does not propose a new and interesting amusement, I shall die of ennui.* She is promptly offered some hashish but declines it, though one of her male companions does take some and subsequently declares his undying love for one of the other female characters; but it is all done in a totally chaste and romantic fashion. Unsurprisingly, there is no record of Alcott having taken drugs.

In Britain, William Butler Yeats, Oscar Wilde, Aubrey Beardsley and Havelock Ellis formed a clique of drug takers and it is difficult to disentangle the effects of hashish on their literary/artistic output from the effects of opium, mescaline and absinthe, which they also consumed. The reprobate and heroin addict Aleister Crowley has already been mentioned in Chapter 2, and his literary output on the subject of hashish was confined to a few essays in which he was mostly disparaging about the drug (though he claimed that it enhanced his sexual performance with both sexes), preferring the stronger effects of heroin, cocaine and mescaline.

Louis Lewin provided his usual accurate and straightforward account (in his *Phantastica*) of his observations, and described how cannabis users proceeded from *an exhibition of gaiety in which smokers behave in a very childish and stupid manner* to a dream-like state where *all the thoughts which pass through the brain are lightened by the sun . . . and the bonds of time and space are broken.*

However, it is the effects of cannabis use on the music scene, especially in the USA, which begins to dominate the social history of this plant from the 1920s. While hemp had been introduced into the Americas by Europeans as a source of fibres for rope and cloth, it was African slaves, shipped in to work in the Brazilian sugar plantations and the American cotton fields, who first used the plant for its euphoriant properties. In the British West Indies, it was Indian workers arriving to replace the slaves in the mid-nineteenth century who brought their *ganja* habit with them. In Jamaica they laid the foundations for the rise of the Rastafarian movement, which was a culture engendered by pan-Africanism and for which *ganja* was a sacred herb for healing and meditation. This led ultimately to the emergence of reggae music in the 1960s, and to its star exponents like Bob Marley, who rose to prominence in the USA and Europe from 1971. Even to this day, many smallholdings in Jamaica will have a small plot dedicated to the growth of *Cannabis sativa*. The use of cannabis spread north through Central America and by the 1880s it was being widely used in Mexico, both with food and in the form of cigarettes, and here it was usually known as *marihuana* (later *marijuana*). Inevitably some cannabis leaked across the border into the southern states of the USA, and it became better known following the incursions of Pancho Villa and his raiding parties from 1916. These brigands were known as 'the cockroaches' and were all users of marijuana – hence the term 'roach' for the stub end of a marijuana cigarette. The song *La cucaracha* also dates from this time:

> *The cockroach, the cockroach,*
> *Now he cannot walk,*
> *Because he don't have, because he's lacking*
> *Marijuana to smoke.*

The surge of marijuana use in cities like New Orleans and Chicago during the 1920s was contemporaneous with the evolution of jazz and blues music, and regular users included such well-known performers as Hoagy Carmichael, Fats Waller and Louis Armstrong. The latter claimed:

> *Marijuana makes you feel good man, makes you forget all the bad things that happen to a negro.*

These musicians gravitated towards the big time music scene in New York, and Harlem in particular, during the 1930s, where so-called 'tea pads' became the place to find both jazz and marijuana. A marijuana joint could be obtained for as little as 15 to 50 cents and the lyrics of the songs of this period became more and more explicit about their source of inspiration. The 'Reefer song' by Fats Waller is a particularly good example of what was known as 'viper music':

Dreamed about a reefer five feet long.
Mighty Mezz, but not too strong.
You'll be high, but not for long
If you're a viper.
I'm the king of everything.
I've got to be high before I can swing.
Light a tea and let it be
If you're a viper.

'Mezz', 'muggles' and 'tea' were local slang for marijuana joints. The first term was derived from one of the great characters of the 1920s and 1930s, Milton 'Mezz' Mezzrow, a jazz clarinet and saxophone player who became in turn a bandleader, entrepreneur and supplier of the best Mexican marijuana, initially in Detroit, then in Chicago and New York City. In his autobiography entitled *Really the Blues*, he reveals that:

During the years when I sold the stuff, I never 'pushed it' . . . I sold
it to grown-up friends of mine . . . overnight I was the most popular
man in Harlem.

Of course these musical activities, enlivened by marijuana and to a degree cocaine and opium too, did not escape the attention of the media. The newspapers of William Randolph Hearst were particularly vehement in the 1930s in their condemnation of the 'evil weed' and the blacks and Mexicans who used it or peddled it. The government was not slow to act, and in 1937 Roosevelt signed the Marihuana Tax Act. This was a strange piece of legislation since it sought to raise revenue rather than criminalising the drug. Every time marijuana changed hands, a tax of $100 per ounce had to be paid and in consequence all those who produced or sold the drug had to be registered. The very open use of marijuana was probably reduced by the Act, but more significantly it had the (apparently) unexpected effect of killing off the farming of hemp, and this was to have serious consequences when America needed rope during the Second World War. Another effect on marijuana use was the appearance of the amphetamines, initially benzedrine and dexedrine, and later methedrine (speed), a class of drugs that were cheap and very readily available.

The musicians were not the only ones to gain notoriety from their drug use, and following the Second World War, the so-called Beat Generation emerged from New York City led by the likes of Allen Ginsberg, William Burroughs and Jack Kerouac. Their literary and other activities were fuelled by a whole medicine cabinet of drugs including morphine, heroin, peyote, cocaine, speed, marijuana and alcohol. Ginsberg is mostly remembered for his political activism (especially against the Vietnam War), and for his explicit poem *Howl*. This was published in 1956, and celebrated the lives of the Beat Generation. It was almost immediately banned for its alleged obscenity. He spent much of his time high on marijuana, peyote or opium, studying the works of William Blake and trying to *expand his mind* and *free his soul*. William Burroughs, an inveterate opium addict, has two claims to notoriety: he shot and killed his common-law wife Joan in 1951 by firing a gun at a glass of water she had balanced on her head – he was of course high as a kite at the time; and he wrote the extremely weird book *Naked Lunch*. This was written over a four-year period in Tangier, partly with the help of Ginsberg and Kerouac, and finally published in 1959 immediately becoming the subject of obscenity trials in several states.

Jack Kerouac made his famous road trip with Neal Cassady (a small-time car thief) in 1948, borne along on a wave of alcohol, speed, peyote and marijuana, and his book of the trip entitled *On the Road* was first published in 1957. By today's standards the book is quite tame with its accounts of criss-crossing America in trucks or wildly driven cars, and descriptions of weird individuals they met along the way, together with attempted seductions of (mainly) teenage girls. But in conservative 1950s America it seemed outrageous. There is little explicit description of drug use, though towards the end the main characters (Sal Paradise aka Jack Kerouac and Dean Moriarty aka Neil Cassady) encounter a woman in Northern Mexico who provided them with some home-grown and very potent marijuana:

> *He proceeded to roll the biggest bomber anybody ever saw. He rolled (using a brown paper bag) what amounted to a tremendous Corona cigar of tea. It was huge.*

They then proceeded to the local brothel to dance and use the local girls and, on the way, Sal looked out of the window and saw Mexico as:

Some gloriously riddled glittering treasure-box . . . . I saw streams of gold pouring through the sky and right across the tattered roof of the poor old car.

And the mambo music in the brothel:

Resounded and flared in the golden, mysterious afternoon like sounds you expect to hear on the last day of the world and the Second Coming.

Neal Cassady took another famous road trip in 1964 when he drove a converted school bus covered in psychedelic art, and fitted with powerful speakers, from California to New York. This 'magic bus' was the brainchild of Ken Kesey, the author of *One Flew Over the Cuckoo's Nest*, and one of the founding members of the Merry Pranksters, who sought to popularise the use of LSD. As the bus travelled across America broadcasting folk rock at full volume, Kesey was on hand at many of its well-publicised stops to supply LSD for experimentation. These exploits formed the basis of the Tom Wolfe novel entitled *The Electric Kool-Aid Acid Test*.

Ginsberg, Burroughs and Kerouac have had a lasting influence on the social history of cannabis and other drugs, and they were the fore-runners in the 1950s of what was to be an explosion of music and lit-erature that emerged in the 1960s – much of it influenced by cannabis and other drugs. The Beat Generation spawned the Beatniks and, especially in the USA, these bearded folk dressed mostly in black clothes, openly used marijuana and spread the word about its effects. Of all the musicians and songwriters who contributed to 1960s culture – The Beatles, The Rolling Stones, Cream, The Eagles, The Beach Boys Led Zepellin and the Who – and who did undoubtedly use drugs, it is Robert Zimmerman, aka Bob Dylan, who epitomises the evolution of folk music in the last 40 years of the twentieth century. He worshipped Woody Guthrie, studied the works of Ginsberg and Kerouac and wrote some of the most (apparently) drug-inspired lyrics of any performer on tracks including *Mr Tambourine Man*, *Like a Rolling Stone* and in particular *Desolation Row*. Dylan has, however, always steadfastly denied taking any drugs apart from marijuana. In his interview for *Playboy*, published in 1966, he was ambiguous to say the least:

I wouldn't advise anybody to use drugs – certainly not the hard drugs. But opium and hash and pot – those things aren't drugs, they just bend your mind a little. I think everybody's mind should be bent once in a while.

At the end of the 1960s, a veritable Who's Who of the folk-rock scene performed at what was probably the biggest musical event of the decade – perhaps of the century: the music festival at Woodstock in upstate New York in August 1969. Close to 500,000 people attended, with the great majority using marijuana at the very least, and spent three days listening to some of the greatest bands of the era. One aspect of this 'happening' that is most striking is the almost total lack of any crime or violence despite the huge number of people in attendance. This contrasts with the alcohol-fuelled problems that are almost inevitably associated with much smaller events in the twenty-first century.

This period of 'peace and love' that started with the 'summer of love' in San Francisco in 1967 spawned the hippie culture, and these bohemian travellers soon moved on from places like Haight-Ashbury and Big Sur in California to Mexico and Morocco where cannabis was widely available. The Arab occupation of Morocco from the eighth century had led to the introduction of cannabis, and from around 1800 a major area of cultivation was in the Rif Mountains. *Cannabis sativa* became a staple agricultural crop for the Berber people and the product made from flower heads and tobacco is known locally as *kif*. Successive rulers of Morocco made little effort to control production of *kif*, until the newly independent government of Morocco used its army in 1962 to try to curb the Berbers' productivity. However, it was beaten back and crop production was unaffected. The hippies had a lasting effect on Morocco, not only effectively taking over cities like Tangier, but their quest for ever stronger forms of cannabis led the Berbers to experiment with more effective ways of extracting the resin from the flower heads, and hashish increasingly displaced *kif* as the extract of choice. The hippies eventually moved on to more exotic locations such as Nepal, India and Afghanistan, and the production of cannabis in each of these countries expanded to meet their needs, though opium production has always been the main activity in Afghanistan.

By the mid 1990s the major producers of hashish were Lebanon (around 20%), Pakistan (30%) and Morocco (35%), with marijuana being exported in large quantities from Colombia, Mexico and Jamaica. The total world production of cannabis in its various forms probably amounted to about 50,000 tons. In all of these countries, peasant farmers could make a good living by growing cannabis as one of their staple crops. In the 1970s, the Mexican border was particularly leaky with marijuana smuggled into California and Texas by car, truck and boat. Much of this was in small amounts brought in by tourists and small-time operators in their private cars, and many Mexican garages specialised in customising cars (door panels, petrol tanks, *etc.*) for the smuggling trade.

Similarly small-scale imports came into Britain from Morocco *via* Spain often concealed within mobile homes and VW caravanettes. Some idea of the scale of these operations and their sheer audacity can be gained from two books, one that celebrates the 1970s activities of the American smuggler Allen Long (*Smokescreen* by Robert Sabbag), and the other an autobiography of the British smuggler Howard Marks (*Mr Nice*). The former flew DC-3s loaded with cannabis from small dirt strips in Colombia into various quiet airports in the USA, and was finally apprehended in 1991 after over 15 years of drug-related activities that included frequent brushes with murderous drug dealers in both Colombia and the US. In return for a guilty plea he received a five-year jail sentence. Howard Marks used less gung-ho methods for his smuggling, and these included commercial flights from Pakistan *via* Shannon (where the Irish police were less vigilant), and thence by ferry across the Irish Sea to Britain. He also used a sea-going tug, sailing from Colombia to the UK, before he was apprehended in 1988 and then served seven years for his crimes.

Needless to say the increasing use of cannabis preparations from the 1920s had attracted the attentions of the legislators, though there was always a sense that cannabis was somehow less dangerous than opium, heroin or cocaine. A major study of marijuana use in New York City was carried out from 1938 to 1944 at the behest of the then mayor Fiorello La Guardia. This involved participation by the inmates of Rikers Island Penitentiary and certain patients in the Memorial Hospital, and had the full co-operation of the New York Police Department. It was claimed to be a comprehensive study of the scientific and sociological aspects of marijuana use, and the final La Guardia report concluded:

> *From the study as a whole, it is concluded that marijuana is not a drug of addiction comparable to morphine, and that if tolerance is acquired, this is of a very limited degree. Furthermore, those who have been smoking marijuana for a period of years showed no mental or physical deterioration which may be attributed to the drug.*

Of course this predated the 1954 report of the British epidemiologists Richard Doll and Richard Peto about the long-term effects of cigarette smoking, and more recent reports of the equally damaging effects of cannabis smoke on the lungs. Despite this apparent clean bill of health, law enforcers in the USA were determined to restrict cannabis use and the Narcotic Control Act of 1956 laid down harsh penalties for possession of heroin, cocaine and marijuana with no discrimination between the three drugs. The Act suggested prison terms of 10, 20 and 40 years for first, second and third offences respectively, with terms of 5–20 years for a first offence of dealing in the drugs. Some US States introduced a mandatory death penalty! The United Nations introduced its own legislation in 1961 with the Single Convention on Narcotic Drugs, with the use of all narcotics and marijuana fully prohibited except for medical uses.

In typical fashion, Allen Ginsberg led one of the first lobby groups to campaign for decriminalisation – LeMar or Legalise Marijuana – which he set up in 1964. He became increasingly vocal and his credibility was finally accepted. In the normally staid *Atlantic Monthly* he wrote a strong article in November 1966 in which he declared:

The actual experience of the smoked herb has been completely clouded by a fog of dirty language by the diminishing crowd of fakers who have not had the experience.

This was coming at the same time as Timothy Leary's exhortation to *turn on, tune in, drop out* (Chapter 1), together with an increased level of social upheaval in the USA (and elsewhere) associated with the Civil Rights movement, student unrest, the Vietnam War, the rise of feminism and the general atmosphere of sex, drugs and rock and roll.

A surge in the use of marijuana in the UK began at the end of the Second World War, most of it smuggled in from Africa or the Middle East. As in the USA the Press began a campaign of fear regarding the evil influence of 'coloured racketeers' peddling the drug to unsuspecting teenage girls. Less seriously, Chapman Pincher in the *Daily Express* promulgated the theory that reefers and rhythm were directly connected *via* the 'minute brainwaves' of jazz fans causing their bodies to gyrate in sympathy with the music. In 1965, the Dangerous Drugs Act placed cannabis and its products in the same category as opiates and cocaine, with a maximum jail sentence of 10 years and a fine of £1000 for possession or trafficking. Several pop stars soon fell foul of this with Mick Jagger and Keith Richards of The Rolling Stones famously arrested at Richards' house in February 1967, allegedly after a tip-off from the *News of the World*. The circumstances of their arrest have provided journalists and scandal mongers with a feast of lurid and far-fetched stories over the years, but the reality is that Richards was found guilty of allowing his house to be used for the smoking of marijuana while Jagger was found guilty of possessing speed (four tablets). The former received a prison sentence of one year and the latter a sentence of three months, eliciting a wonderful editorial from William

Rees-Mogg of *The Times* entitled *Who breaks a butterfly on a wheel* in which he wrote:

There must remain a suspicion in this case that Mr Jagger received a more severe sentence than would have been thought proper for any purely anonymous young man.

Both performers were eventually freed on appeal, though Jagger only received a conditional discharge. They were the first of a string of famous offenders charged with possession of marijuana, including Paul McCartney, John Lennon and Bob Marley. Certainly the pop music industry of the 1960s, and probably the film industry too, seems to have been borne along on a wave of marijuana. Richard Lester, director of The Beatles' film *Help*, is reputed to have claimed that the aroma of cannabis smoke pervaded the set at all times. The many on-screen antics of the stars would certainly support this assertion.

These arrests triggered widespread demand for law reform and famously led to a newspaper advertisement placed in *The Times* by the group SOMA (Society of Mental Awareness). Sixty-five liberals and freethinkers, including Francis Crick (of DNA fame), the critic Kenneth Tynan, Graham Greene and the photographer David Bailey, signed a statement that declared *the law against cannabis is immoral in principle and unworkable in practice.*

The newspaper comment accompanying these famous arrests and this advertisement caused the UK government to review the dangers of marijuana, and Baroness Wootton, a well-respected sociologist, was commissioned to investigate these matters. In the event her report of January 1969 made uncomfortable reading for a government intent on establishing the dangers of cannabis use. Interestingly the report cited not only the La Guardia Report but also the original Indian Hemp Drugs Commission of 1894. This had heard evidence from more than 1000 witnesses on the use of hemp by all sections of the Indian community. Typical was the statement made by Colonel Alexander Crombie, one-time surgeon superintendent at Calcutta Hospital, who claimed that:

My servant who uses ganja or bhang in excess, is always at his post and capable of doing his duty.

The Commission's chairman, Sir William Mackworth Young, had concluded:

On the whole the weight of evidence is to the effect that moderation in the use of the hemp drugs is not injurious to health.

Having considered all of the evidence, Baroness Wootton concluded:

Long-term consumption of cannabis in moderate doses has no harmful effects . . . and is very much less dangerous than the opiates, amphetamines, and barbiturates, and also less dangerous than alcohol.

She also made two recommendations for some easing of the restrictions on cannabis:

Possession of a small amount of cannabis should not normally be regarded as a serious crime to be punished by imprisonment.

And:

Preparations of cannabis and its derivatives should continue to be available on prescription for purposes of medical treatment.

The Government promptly ignored these liberalising aspects of her report and passed the Misuse of Drugs Act in 1971 which divided drugs into three classes of increasing danger. Class A included morphine, heroin, LSD and cocaine; class B included amphetamines, cannabis products and barbiturates; while class C was a miscellany of medical products that were in principle abusable.

Interestingly, at about the same time, the US Government introduced the Comprehensive Drug Abuse Prevention and Control Act (in 1970), which downgraded possession of cannabis products to a misdemeanour. The Canadian Government followed suit in 1972. President Jimmy Carter was probably the most liberal in his views on marijuana, and in an address to Congress in August 1978 he warned against severe penalties for possession of a drug that were more harmful than the drug itself:

Nowhere is this more clear than in the law against possession of marijuana in private for personal use.

However, at the present time, importation of the drug into the USA can attract a life sentence or even the death penalty in some states. Similar severe penalties still exist in Malaysia and Singapore.

Since then there have been a number of attempts to decriminalise the drug in the UK. The Runciman Report of 2000 suggested cannabis products should be reclassified to class C, and while David Blunkett was Home Secretary this reclassification did occur (in 2002) with the police gradually taking a more tolerant attitude to the possession of small quantities of cannabis for personal and private use. In particular, the growing awareness that the medicinal properties of cannabis were more substantial than had been realised encouraged this more enlightened view. However, this was not to last for very long, due to the rise in the number of 'cannabis farms' growing the resin-rich form of cannabis known as *skunk*. These 'farms' range in size from domestic garages to full-size nurseries with full hydroponic facilities. The Government's response has been to introduce legislation that includes a reclassification of cannabis to category B from 2009.

The medicinal properties of cannabis have been known for millennia. As well as the documented use by the great Persian physician Rhazes, the famous English herbalist Culpeper in his *Complete Herbal* (1653) sings the praises of hemp for allaying *troublesome humours of the bowels* and *inflammation of the head, and other parts*, and also claims its value in the treatment of gout, jaundice and worms. But it was in Victorian times that preparations like Squire's Extract and Dr J. Collis Browne's Chlorodyne (already mentioned in Chapter 2) became popular for the treatment of everything from incontinence, neuralgia and infantile convulsions to anthrax, cholera and rabies. Both of these proprietary preparations contained extracts of hashish in alcohol, though the Chlorodyne contained quantities of morphine and chloroform as well. The Edinburgh-educated Irish physician Dr William O'Shaughnessy observed the use of Indian hemp while working in India, and popularised the use of tincture of cannabis resin in London on his return. He advocated it especially as an anti-convulsive remedy, though he did note that injudicious use could lead to *a strange balancing gait and perpetual giggling*. Various other Victorian physicians noted its value in the treatment of chronic headache, for example C. R. Marshall in 1905 claimed: *It appears to be useful in headaches of a dull, continuous character.*

A brief foray into the isolation of cannabinoids took place in 1896, but it was not until 1940 that Alexander Todd and his group at Cambridge and Roger Adams at the University of Illinois isolated and provided a structure for cannabinol, an inactive constituent of the plant. It took a further 23 years before Mechoulam and Shivo at the Weissman Institute in Israel provided a structure for cannabidiol – also inactive – and finally the major psychoactive constituent Δ^9-tetrahydrocannabinol succumbed to the structural investigations of Mechoulam and Gaoni, also at the Weissman Institute in Rohovot, Israel, in 1964. Altogether nearly 500 discrete compounds have been isolated from *Cannabis sativa*, but only Δ^9-THC has any appreciable psychoactive activity. One of the most interesting of the other cannabinoids is Δ^9-tetrahydrocannabinolic acid since this appears to be a major constituent in the plant but loses a CO_2 group (decarboxylation) to yield Δ^9-THC if cannabis products are smoked or eaten. Mechoulam and Gaoni were also responsible for the first chemical synthesis of Δ^9-THC in 1967, and since then a whole range of natural cannabinoids and analogues have been prepared, but most have proved to be inactive in psychoactivity studies.

One analogue that has begun to achieve some notoriety first appeared in the laboratory of John W. Huffman of Clemson University in 1995, when he synthesised the compound N-butyl-3-(1-naphthoyl)indole which he christened JWH-018 for short. He wanted to use this compound to study structure-receptor relationships at cannabinoid receptors (see below). When tested, and despite the wide difference between this structure and Δ^9-THC, the compound was about 4–5 times more potent than THC and bound to both CB1 and CB2 receptors. It is also relatively easy to prepare and has already attracted the attentions of the 'bathroom chemists' who are selling JWH-018 in various herbal preparations. These go under such names as Spice Gold and Spice Yucatan Fire, and are beginning to be very popular. Not surprisingly governments have begun to move towards making these products illegal.

As with opium and its products, the obvious question every investigator wanted to answer was: why does this unusual structure have psychoactive properties? A partial answer came in 1988 with the discovery of mammalian cannabinoid receptors, that is proteins within the body that interact with Δ^9-THC and trigger off a response. In the event, two types of receptors were identified, so called CB1 and CB2, with the former most prevalent in the brain, and to a lesser extent in the lungs, liver and kidneys, while the latter is mainly confined to peripheral cells of the immune system and the spleen. Once activated there is a triggering of a number of cell signalling pathways together with involvement of potassium and calcium channels (in the cell membranes). The reduction in calcium ion uptake may explain the inhibition of acetyl choline release in the area of the brain called the hippocampus. It also leads to a reduction in the release of the neurotransmitter noradrenaline at sympathetic nerve terminals, and in the hippocampus, the cerebral cortex and the cerebellum. These changes in neurotransmitter levels probably explain the euphoriant activities and also the changes in cognition and memory. Interestingly, activation of CB1 receptors in the liver increases fat production and at least one of the new anti-obesity drugs – ribonabant (Acomplia) – acts by inhibiting CB1 receptor activation.

Of course this was only part of the answer, since it still begged the question – why do mammals have receptors that interact with a chemical compound from *Cannabis sativa*? The answer came in 1992, again from Mechoulam's research group, with the isolation of a fatty acid derivative – arachidonylethanolamide – which also bound to cannabinoid receptors and elicited similar effects to those produced by Δ^9-THC. They christened this anandamide, from the Sanskrit *ananda* meaning bliss or delight, and it proved to be the first of eight endogenous cannabinoids. They appear to be involved in pain control, in regulation of feeding behaviour, in control of the pleasure response, in memory and in a whole host of other neural responses. For a while in the late 1990s it seemed they might also be responsible for the all too guilty pleasures associated with chocolate consumption. A paper in *Nature* in 1996 claimed that anandamide was present in chocolate at very low levels accompanied by analogous structures – oleoylethanolamide and linoleoylethanolamide. These last two compounds had already been prepared and shown not to interact with cannabinoid receptors, and later studies failed to confirm the presence of anandamide in chocolate. So the neurological basis of chocolate's appeal still remains elusive. That said, oleoylethanolamide is an endogenous regulator of food intake and thus has natural anti-obesity effects, while the structurally related oleamide is an endogenous sleep-inducing factor.

Given the presence of cannabinoid receptors in the body, cannabis use has been investigated for a number of medical conditions over the past ten years or so. Its anti-emetic effect has been known since the days of the hippies in the early 1970s, and both Δ^9-THC and a synthetic analogue called nabilone have been used in conjunction with cancer chemotherapy. In fact the synthetic analogue nabilone is more effective than many other anti-emetic agents and also appears to enhance the appetites of cancer patients and AIDS patients. Δ^9-THC has also shown anti-proliferative activity with certain types of brain tumour cells (gliomas) and prostate cancer cells, but the mechanism of action has not yet been determined.

In addition, the use of cannabis in conjunction with more conventional drugs has been shown to alleviate some of the symptoms of multiple sclerosis and arthritis, and the British Medical Association recommended legalisation of cannabis for these purposes in 1997. Courts in the UK and USA have proved unsympathetic and have thus far rejected this proposal, while the governments of the Netherlands, New South Wales and Canada have relaxed the rules governing cultivation of cannabis for personal medical use. This situation may change if the UK company GB Pharmaceuticals gains approval for its oral spray called Sativex (a mixture of Δ^9-THC and tetrahydrocannabinol) for alleviation of the symptoms of MS. Of course the trick would be to produce an analogue that bound to cannabinoid receptors and produced analgetic and other useful effects without inducing euphoria *etc*. Unfortunately this feat has not yet been accomplished, though for those who suffer from various kinds of neuropathic pain not easily treated with regular analgesics, the kind of experience described by Baudelaire in *Les Paradis Artificiels* would probably be welcome:

> *Le temps avait complètement disparu. Tout à l'heure c'était la nuit, maintenant c'est le jour* (in a minute it was night, now it's daytime) . . . *la nuit entière n'a-t-elle eu pour moi à peine que la valeur d'une seconde* (the whole night was little more than a second).

CHAPTER 5

Belladonna, Mandrake and Daturas

*The witches confess, that on certain days or nights they anoint a
staff and ride on it to the appointed place.*

This report from the trial of an alleged witch in 1470 is pretty much in
line with most of the documentation from European sources for the
fifteenth to seventeenth centuries. The confession of the Belgian witch
Claire Goessen in 1603 provides further exemplification:

*Elle s'est laissé transporter au moyen d'un baton enduit d'onguent (a
stick anointed with unguent).*

Two major questions need to be addressed: who were the witches and
what was the nature of the unguent or witches' salve? There are two
extreme answers to the first question and you can suspend belief
and soak up the occultism presented by Montague Summers in his
well-known book *History of Witchcraft and Demonology* (1928). Alter-
natively, you can consider the much less extreme views of the well-
respected Egyptologist Margaret Murray in her book *The Witch-cult in
Western Europe* (first published by Oxford University Press in 1921). She
proposed that witchcraft was a pre-Christian cult that revered Nature in
all its forms with a special emphasis on fertility, and with a pagan deity
who was probably bisexual. Although her views attracted considerable
controversy, it is certainly true that European magic and sorcery had

Turn On and Tune In: Psychedelics, Narcotics and Euphoriants
By John Mann
© John Mann 2009
Published by the Royal Society of Chemistry, www.rsc.org

their roots in Graeco-Roman pagan cults and these were later complemented by Viking, Norse and Celtic folk practices. All of these incorporated the Roman goddess Diana (or an equivalent deity), who had a strong association with what was called the Wild Hunt. This was depicted as a wild orgiastic night-time hunt involving animals, wraiths and spirits. The great pagan festivals of Walpurgis (evening of 30th April), Midsummer's Eve (20th June, the eve of the summer solstice) and All Hallows Eve (31st October, the last day of the pagan year) also involved wild orgiastic celebrations and bonfires. Together these pagan rituals provided the basis for the witches' Sabbat, where the witches were said to dance around a man dressed in black (*le diable*) or *un grand chat noir* and, while couplings of various sorts were claimed to occur, most written reports also include statements like *va baiser le derrière du diable*.

So who were the women who became involved in these weird activities? First of all it is important to note that in the fifteenth to seventeenth centuries women did most of the food preparation and cooking, and were also healers and midwives. In consequence any problems with food or beer becoming tainted, or women dying in childbirth, or folk medicines killing the patients – in fact any kind of bad luck – could be laid at their doors. Midwives, in particular, came under increasing suspicion during this period, and it was often claimed that these women would offer newborn babies to the Devil during special post-delivery baptisms. From around 1450 a witch craze gripped Europe for almost 200 years, exacerbated by the great conflict between the Protestant and Catholic religions, and women became the major target for this paranoia. The infamous publication *Malleus Maleficarum* (*Hammer of the Witches*), written by the Dominican friars Heinrich Kramer and Jacob Sprenger, who were also participants in the Inquisition, was published in 1486. This encouraged a belief in magic and witchcraft of all kinds and, although banned by the Catholic Church, it served as a manual for witch-hunting throughout the sixteenth and seventeenth centuries. Amongst other things it claimed that witches were usually women since women were more stupid, weaker and more carnal than men, and that they changed their shapes, flew through the air and concocted magic ointments. Similar claims were made in *De Secretus Mulierum (Women's Secrets)*:

> *Women are so full of venom in their time of menstruation that they poison animals by their glance, they infect children in the cradle; and whenever men have sexual intercourse with them, they are made leprous and sometimes cancerous.*

Given this climate of fear and suspicion, it is hardly surprising that throughout Europe thousands of women were arrested and confessions were wrung from them through use of the thumb screw, water torture, strappado and the rack. The British Isles were spared the very worst excesses of the Inquisition and witch trials were relatively few in number. One of the most famous trials followed accusations against Alice Kyteler in 1324, an alleged witch from Kilkenny. It represents one of the first documented cases where an ointment was mentioned:

In rifleing the closet of the ladie, they found a pipe of ointment, wherewith she greased a staffe, upon which she ambled and galloped through thick and thin.

The full list of charges included associations with demons, the practice of various sorceries, the making of potions in the skull of a decapitated felon and using sorcery to kill or enfeeble each of her four husbands. In the event, she was never brought to trial since she escaped to England leaving several of her associates to face her accusers. One of these, Petronella of Meath, confessed under torture to sorcery and was burned at the stake.

There is overwhelming support for the belief that witches rubbed their bodies with an ointment or salve which led to levitation using animals, broomsticks, stools or pitchforks. Giambattista della Porta, a friend of Galileo, described how the witches rubbed ointment on their bodies:

They anoint the parts of the body, having rubbed them very thoroughly before, so that they grow rosy . . . and they think they are carried off to banquets, music, dances, and coupling with young men.

And a similar story was told (in 1584) by Reginald Scot who identified some of the ingredients of the ointment:

Blood of a flitter mouse, solanum somniferum, and oleum . . . they rubbe all parts of their bodys exceedinglie, till they looke red, and be verie hot, so as the pores may be opened, and their flesh soluble and loose . . . by this means in a moonlit night they seeme to be carried in the aire.

From a very long catalogue of alleged constituents including soot and aconite (unlikely due to its great toxicity), the three famous hexing herbs emerge from most recipes. The hexing herbs were deadly nightshade (*Atropa belladonna*, see Figure 5.1), black henbane (*Hyoscyamus niger*)

Figure 5.1   A photo of *Atropa belladonna*. © Jan Samanek, State Phytosanitary
Administration, Bugwood.org.

and mandrake (*Mandragora officinarum*). All of these contain tropane
alkaloids, most importantly atropine and hyoscine (scopolamine), with
the latter as the most probable facilitator of (apparent) levitation.
Scopolamine is a powerful elicitor of hallucinations especially involving
shape and vision, and leads ultimately to narcosis, while atropine will
cause disorientation and hallucinations but only in toxic doses leading to
respiratory failure, paralysis and death.

If you were a witch, the trick was to get the scopolamine into the brain
without exceeding the toxic dose. The alkaloid is absorbed rapidly from
the gastrointestinal tract but it is also easily absorbed through mucosal
surfaces – the armpits and genital region – and here the anointed
broomstick would have played a key role. In addition, the use of fat in
the ointment and the rubbing of the skin until it was red would have
ensured uptake through the skin (see Figure 5.2). Once in the bloodstream, scopolamine penetrates the blood-brain barrier very effectively
(unlike atropine) and then causes depression of the central nervous
system through antagonism of the actions of the neurotransmitter
acetylcholine. An initial feeling of drowsiness and amnesia would
have been followed by euphoria, hallucinations and finally narcosis.

DEPART POUR   LE SABAT

Grav.º d'après le tableau Original de D. Teniers   du Cabinet de Monsieur le Comte de Vence.

à Paris chez Alliamet graveur rue des Mathurins la 4.º   Porte cochere   gauche en entrant par la rue de la Harpe.

Figure 5.2 A witch preparing for the Sabbat – note the unguent being prepared in the foreground and the witch astride her broomstick with unguent being applied by a helper. © Wellcome Images.

These responses will be familiar to anyone who received premedication prior to surgery, back in the days when the NHS took a more relaxed view of surgical procedures and was less concerned about the turn around of patients. These premedications typically contained scopolamine and opiates and it is the closest that most people will have come to experiencing the effects of a witch's salve.

Of the other hexing herbs, the mandrake root has the most interesting folk history, and various preparations have been used for at least 5000 years. Some of the mystique surrounding the plant undoubtedly arises from the appearance of the root. This is often shaped like the letter Y, and thus may be compared with the human form, so opening the opportunity for use in magic, though it was also valued as an adjunct to surgery and also as a poison. Not surprisingly, the magicians, herbalists and apothecaries dreamt up ever more imaginative means of ensuring that this valuable commodity did not fall into the wrong hands. The Greek herbalist Theophrastus (third century BC) was amongst the first to describe a collection procedure involving a circle cut around the plant with a sword prior to uprooting while facing towards the west. Pliny the Elder (first century AD) also wrote of the foul stench produced during the uprooting. The Jewish historian Josephus (first century AD) provided more details including the use of a dog tied to the root to assist with the uprooting. Most later descriptions, including a large number of murals and woodcuts, also mention or depict a dog tied to the root (see Figure 5.3), but add something about the dreadful shriek emanating from the root at the moment of removal from the soil. This awful sound was alleged to kill the dog and render mad anyone who heard it. Hence:

> *Then on the still night air,*
> *The bark of a dog is heard,*
> *A shriek! A groan!*
> *A human cry. A trumpet sound,*
> *The mandrake root lies captive on the ground.*

And while the author of that particular verse is not known, the references in Shakespeare to the same events are well documented as when Juliet says (in *Romeo and Juliet*):

> *Shrieks like mandrakes, torn out of the ground,*
> *That living mortals, hearing them, run mad.*

Figure 5.3 A woodcut depicting mandrake collection. © The Bridgeman Art Library/Getty Images.

Now J. K. Rowling has brought this myth to a wider audience and this was depicted most graphically in the second Harry Potter film (*Harry Potter and the Chamber of Secrets*), where the pupils in the practical class being taught by Professor Sprout put on their earmuffs before uprooting their mandrakes.

The beneficial medical uses were described by William Turner in his *New Herball*, which appeared in three parts between 1551 and 1568, and he suggested that:

If mandragora be taken out of measure by and by slepe ensueth and a
great lousing of the strygthe with a forgetfulness.

Shakespeare confirmed the soporific effect of mandrake in *Anthony
and Cleopatra* [Act 1, Scene 5]:

Give me to drink mandragora,
That I might sleep out this great gap of time
My Anthony is away.

The Renaissance poisoners like Lucrezia Borgia are alleged to
have fermented the root until it stank, then stored it for a further 60 days
until a greenish pulp was obtained. This was administered to the
victim, who then suffered trembling, stomach pains and yellowing of
the eyes before they died. If the victim suspected that this was
something a bit worse than a dose of dysentery or the effects of too much
alcohol, there was apparently an antidote. This, however, comprised
honey, butter, oatmeal, mint leaves, anise, nutmeg, fennel seed, clove
bark, ginger, cinnamon, celery, radish, dill *etc.*, and while these ingredients would be easily found in the modern kitchens of those addicted
to the television chefs, they would have been less common in the
average Renaissance kitchen. The victim would have to seek help from
the local apothecary, who might well have made up the poison in the
first place!

The twentieth-century poisoner Dr Crippen was able to obtain his
hyoscine (scopolamine) from a pharmacy in New Oxford Street. He used
this to poison his wife Cora, though he also shot her for good measure.
Although he subsequently cut up her body and either burnt the body
parts or destroyed them with quicklime, he was careless enough to leave
a few parts in the basement of his house. He was eventually apprehended
and convicted because traces of scopolamine were detected in fragments
of Cora's stomach, kidneys, liver and intestines.

All of these activities took place in Europe or the Middle East using
the classical hexing herbs, but in the Americas plants containing the
tropane alkaloids, especially scopolamine, were also used for magic and
divination. Probably the most important of these plants were those of
the *Brugmansia* sub-genus especially *B. arborea*, *B. aurea*, *B. sanguinea*,
B. suaveolens and *B. versicolor*. All of these have large, bell-shaped
flowers, and seeds from these plants were ground to a powder and made

into a beverage which was much prized for its divinatory properties. The following is a report from an unnamed traveller in Peru in 1846:

The native fell into a heavy stupor, his eyes vacantly fixed on the ground, his mouth convulsively closed, and his nostrils dilated. In the course of a quarter of an hour, his eyes began to roll, foam issued from his mouth, and his whole body was agitated by frightful convulsions.

Plants of the sub-genus *Brugmansia* are closely related to the many *Daturas* which are widely distributed throughout the world, and also contain varying amounts of atropine and scopolamine. The great Persian physician Avicenna extolled the virtues of *D. metel* for medicinal purposes, and it is still used in Asia (often in admixture with cannabis and tobacco) for fevers, skin diseases and analgesia. Similar medicinal uses were claimed for *D. inoxia* and *D. fatuosa* in South America by various Indian tribes, though the plants were also used for divination and for infanticide – breastfeeding mothers were known to smear their breasts with plant extract to kill unwanted babies. Most infamous of all was *Datura stramonium* (see Figure 5.4) also known as devil's trumpet, thornapple and jimson weed, which was primarily used for

Figure 5.4 A photo of *Datura stramonium*. © Howard F. Schwartz, Colorado State University, Bugwood.org.

seduction and induction of temporary madness. The name jimson weed derives from the term Jamestown weed and probably relates to an occasion when British soldiers were drugged by the settlers in Virginia to prevent them apprehending some local renegades (in 1676). Several descriptions of the effects of this scopolamine intoxication were recorded at the time:

> *Some of them having eaten plentifully became fools for several days, one would blow up a feather in the air, another sit naked like a monkey, or fondly kiss and paw his companions, and sneer in their faces.*

To this day extracts of *Datura stramonium* are used by thieves in South American cities to drug their victims so that on awakening they have no recollection of what had befallen them. On a more positive note, cigarettes made from *Datura stramonium*, cannabis and lobelia have been used to treat asthma, and some benefit may be obtained due to the relaxation of bronchial smooth muscle (and resultant bronchodilation) produced by the constituents in the smoke. The drug ipratropium bromide (Atrovent) is a synthetic analogue of atropine and is used in the treatment of chronic obstructive pulmonary disease.

The most instructive and most quoted modern description of *Datura*-induced intoxication and hallucination was provided by the Mexican Carlos Castaneda, who experimented with 'flying ointment' given to him by Yaqui Indians of Northern Mexico in 1968. He reported:

> *The motion of my body was slow and shaky . . . . The momentum carried me forward one more step, which was even more elastic and longer than the preceding one. And from there I soared . . . I saw the dark sky above me, and the clouds going by me . . . I enjoyed such freedom and swiftness as I had never known before.*

However, more recently many of his reports of his experiences have been called into question as works of fiction.

On the other side of the world in Australia, plants of the family *Duboisia* are also a rich source of scopolamine. The species *Duboisia hopwoodii* has always been especially prized by the Aborigines and extracts (known as *pituri*) are used primarily to alleviate fatigue. In the past the Aborigines chewed the plant much as the South Americans Indians chewed coca, but now they smoke it in admixture with tobacco. Joseph King, who was one of the few survivors of the ill-fated Burke and Wills expedition that attempted to cross the central deserts of Australia, recalled how he was found by Aborigines and given *pituri* to help him revive.

The final hexing herb was henbane and as far back as the time of Pliny the Elder it was being recommended for its sedative effect:

> *It is certainly known, that if one take of it in drink more than four leaves, it will put him beside himself.*

A more recent account of the effects of inhaling the smoke from smouldering seeds of henbane was given by Gustave Schenk in 1966:

> *Each part of my body seemed to be going off on its own, and I was seized with the fear that I was falling apart. At the same time, I experienced an intoxicating sensation of flying.*

We will never know exactly what happened at the witches' Sabbats and how much of what they related of their activities had been wrung from them under torture, but the effects of the tropane alkaloids were eloquently described by Louis Lewin. In his *Phantastica* he states:

> *The disagreeable symptoms of these Solanaceae and their active elements, atropine and scopolamine that give rise to hallucinations and illusions of sight, hearing and taste.*

And he was in no doubt about the malevolent effects of these compounds:

> *These substances in suitable doses may give rise to a state of dementia lasting for hours and even days. I am certain that it has frequently been employed for purely criminal purposes . . . and have served to intoxicate girls and seduce them to immoral acts.*

If only the witches had known a bit more about the pharmacology of the drugs they were taking, they might well have thought twice before using them, or at least been able to mount a credible defence at their trials!

CHAPTER 6

Peyote and Amphetamines

There is another herb . . . of the earth. It is called peiotl. It is white.
It is found in the North Country. Those who eat or drink it see
visions, either frightful or laughable (visions espantosas o irrisibles).

Once again the Franciscan missionary Bernadino de Sahagun provides
us with the first and most precise description of a South American
hallucinogenic preparation, in this instance the one from the peyote
cactus *Lophophora williamsii*. This entry comes from his *Historia General de las Cosas de Nueva España* first published in 1557, which also
includes descriptions of *ololiuqui* and other magical preparations. The
grey-green, spineless cactus (see Figure 6.1) is commonly found in
Central Mexico and parts of the south-western USA, and there is evidence from meso-American caves in Texas that early American cultures
may have used this cactus more than 5000 years ago. There are certainly
reports by Francisco Hernandez, who travelled with Cortes, of its use by
the Aztecs as a divinatory plant. In the 1950s, the ethnobotanist Richard
Evans Schultes met several shamans who described to him the importance of the cactus for the Huichol people of central Mexico. They still
have an annual ceremonial hunt for the cactus during the autumn
followed by a kind of harvest festival in January when the peyote is
consumed. Schultes noted that the rituals observed prior to gathering
the cactus were reminiscent of those associated with the Eleusinian
festival – that is confession, sexual abstinence and ritual washing.

Turn On and Tune In: Psychedelics, Narcotics and Euphoriants
By John Mann
Published by the Royal Society of Chemistry, www.rsc.org

Figure 6.1 Peyote cactus – source of the Aztec magical preparation *peiotl* – which contains mescaline as its major hallucinogen. © iStockPhotos.

He also recorded how after consumption of the peyote, the intoxicated participants spent several days dancing and chanting, and offering prayers to their Gods.

All of these cultures used the crown of the cactus to produce slices, now called mescal buttons, and four to thirty of these were consumed as part of pseudo-religious celebrations. From about 1690, the Amerindians of Mexico elevated these celebrations to a form of Mass or Eucharist where mescal buttons were consumed in place of the 'host', activities that did not endear them to the Catholic Church. The American Indians from further north, primarily Mescaleros Apaches, Commanches and Kiowas, appear to have discovered the use of peyote whilst on raiding missions, and brought the custom back to their tribal lands. Here a new religious cult developed from about 1880 and this became known as the Native American Church. This Church grew in popularity and numbered more than 13,000 worshipers by the 1920s. Today the number of active adherents is in excess of one million, many of whom are Navajo Indians. Various attempts were made over the years by the US Government to eradicate this cult with claims that consumption of peyote led to licentious and drunken behaviour. During several well-publicised court battles, the Indians were able to prove that in contrast to popular opinion, peyote use encouraged abstemious behaviour and allowed communication with God rather than anti-social behaviour. Finally in 1967, the US Congress voted to allow the use of peyote by adherents of the Native American Church. This was followed by the Religious Freedom Restoration Act of 1993, which resulted from a Supreme Court ruling in 1990 in favour of two Native American Indians who had been fired from their jobs after testing positive for mescaline. They claimed they had used peyote as part of religious devotions. These days it is possible to obtain mescal buttons by secure mail-order using the US Postal Service.

Over the years a number of well-known authors have written about their experiences with peyote, and most agree that after initial ingestion there are unpleasant effects including nausea and vomiting, and perhaps anxiety attacks, but this is followed by coloured visions, heightened perception of sound and smell and a sense of well-being. The effective dose is around a quarter to a half gram of mescaline and the psychoactive experience may last for up to 12 hours.

One of the first to describe the effects in detail was Louis Lewin in his all encompassing book *Phantastica: Narcotic and Stimulating Drugs*. In this he talks about his first experiments with peyote which commenced in 1888:

Coloured arabesques and figures appear in endless play . . . geometrical forms of all kinds, spheres and cubes rapidly changing colour, triangles with yellow dots from which emanate golden and silver strings, radiant tapestries, carpets, filigree lacework in blue, brilliant red, green, blue and yellow stripes . . . trees with light-yellow blossoms, and many things besides.

And he quite reasonably suggests that in the Indian mind this would be an evocation or personification of God.

Probably the most famous user was Aldous Huxley, who described his experiences in the book *The Doors of Perception*, published in 1954. His main contention was that survival of an individual required what he called the *mind at large* to be funnelled through *the reducing valve of the brain and nervous system* and that mescaline removed these constraints. However, he believed that this occurred as a result of a diminished supply of glucose to the brain rather than (as we now know) due to changes in the levels of neurotransmitters. So an experience with mescaline allowed the consumer to be:

Shaken out of the ruts of ordinary perception, to be shown for a few timeless hours, the outer and the inner world . . . as they are apprehended . . . by the Mind at Large.

Huxley consumed 0.4 g of mescaline in May 1953 and reported the usual array of visual illusions such that a flower arrangement became:

what Adam had seen on the morning of his creation – the miracle, moment by moment, of naked existence.

And a self-portrait of the artist Cézanne in a book that he had in front of him came to life:

as a small goblin-like man looking out through a window in the page before me.

These initial experiences must have been especially exciting for Huxley since he had suffered from seriously impaired vision following a staphylococcal infection when he was 17. Huxley went on to write a second book entitled *Heaven and Hell* in which he further examined the benefits and importance of visionary experiences and concluded that the mescaline experience:

Throws light on the hitherto unknown regions of his (the consumer's) own mind; and at the same time it throws light, indirectly, on other minds, more richly gifted in respect of vision than his own . . . he comes to a new and better understanding of the ways those other minds perceive and feel and think.

On the day that he died, Huxley used LSD to help him achieve a gentle passing and was certainly unaware that earlier that day (22nd November 1963) John F. Kennedy had been assassinated in Dallas. Huxley is, of course, remembered most for his many novels including most notably *Brave New World*, but *The Doors of Perception* remains one of the most significant practical and philosophical treatises on the use of drugs for aiding introspection. It is also famous, so it is alleged, as the inspiration for the name used by the rock band 'The Doors'. Also, like Timothy Leary and Aleister Crowley, Huxley appears on the sleeve of The Beatles' *Sgt. Pepper* album.

Rather similar experiences were reported by Havelock Ellis, a British physician and social commentator who was probably the first to write openly about homosexuality and many other aspects of sexual relations in his hugely influential series entitled *Studies in the Psychology of Sex*. His own marriage was unusual since his wife Edith was a lesbian, and he himself was allegedly impotent though he could apparently be aroused by the sight of a woman urinating. His many books and essays on the subject of sex were complemented by several books of poems and a translation of Zola's *Germinal*. It is perhaps not surprising that such a polymath would experiment with drugs and he wrote of his experiences (in the *Contemporary Review*) after consuming three mescal buttons on Good Friday 1897:

> *In the course of the evening, they (images) became distinct . . .*
> *mostly a vast field of golden jewels, studded with red and green*
> *stones, ever changing . . . they would spring into flower-like forms*
> *and then turn into gorgeous butterfly forms or endless folds of glis-*
> *tening, iridescent, fibrous wings of wonderful insects.*

Perhaps of more interest was his comparison of the effects of the various drugs he had tried:

> *Under the influence of alcohol, the intellect is impaired; haschisch*
> *again produces an uncontrollable tendency to movement and bathes*
> *its victim in a sea of emotion. The mescal drinker remains calm*
> *and collected amid the sensory turmoil around him . . . It may be*
> *claimed that the artificial paradise of mescal is safe and dignified*
> *beyond its peers.*

Finally, a strange little book appeared in 1972 entitled *Misérable Miracle* written by the Frenchman Henri Michaux. In this he describes his numerous experiences of mescaline and compares it with hashish, and the book also includes 48 rather peculiar and not at all artistic drawings. His comparison of drawings carried out under the influence of mescaline clearly had significance for him even though the actual drawings are rather childlike:

> *Les dessins que je faisais après la mesaclin . . . étaient faits*
> *d'innombrables lignes fines, paralléles, serées les unes contre les*
> *autres avec un axe de symétrie principal et des répétitions sans fin.*
> *Très différents, les dessins que je faisais après le haschisch étaient*
> *gauches, embarrassés, morcelés, interrompus prématurément.*

These very personal literary excursions using the natural product mescaline have to be compared with the very public and large-scale use and abuse of the amphetamines, which are chemically synthesised structural analogues of mescaline. The story begins with the investigations in the early 1920s on the Chinese herbal medicine known as *Ma huang* from the plants *Ephedra sinaica* and *Ephedra vulgaris*.

These extracts had been in continuous use since about 3000 BC for the relief of coughing and fever. In the nineteenth century the extracts were known colloquially as *teamster's tea* and *whorehouse tea*, to identify their alleged activities as stimulants for truck drivers and brothel patrons respectively. Pharmacological research carried out in 1923 by Ku Kuei Chen at the Peking Union Medical College led to the isolation of a crystalline alkaloid from *Ephedra sinaica* which proved to be ephedrine, already isolated by Nagajosi Nagai of Tokyo University in 1887, who also established the correct chemical structure. Intravenous injection of ephedrine into dogs produced similar effects to those elicited by adrenaline (called epinephrine in the USA), that is increased blood pressure and accelerated heart rate. Further biological studies in the USA revealed that ephedrine produced a sustained bronchodilator effect in the lungs, and this had great potential for the treatment of asthma. It was approved for this purpose by the American Medical Association in 1926. However, a scarcity of plant material led to a programme of synthesis including analogue studies by Gordon Alles at the University of California in San Francisco and also by researchers at the pharmaceutical company Smith, Kline and French. One particular analogue, Benzedrine, was shown to be an excellent nasal decongestant and was sold in the form of the Benzedrine inhaler from 1932. This simple racemic phenylpropylamine was also called *amphetamine*, and had first been synthesised in 1887 by Lazar Edeleanu at the University of Berlin. It quickly became apparent that it was a powerful stimulant in the central nervous system as well as a decongestant and, during the Second World War, Benzedrine and the more potent (S)-methylamphetamine (Methedrine, 'speed'), first synthesised in 1929, and dexamphetamine (Dexedrine), one of the pure stereoisomers of amphetamine, were used by the armed forces to alleviate fatigue. They were also much employed in the Korean and Vietnam Wars.

After the Second World War the amphetamines became widely available as prescription drugs for the treatment of obesity, post-natal depression and stress and, in some countries like Japan, they were considered safe enough to be purchased over the counter. This combination of ready availability and the emerging drugs culture of the 1960s almost inevitably led to the amphetamines becoming yet another agent of abuse. Intravenous use of methedrine was a particularly prevalent and dangerous practice, and hence the slogan 'speed kills'. In the spring of 1969, the body formed between the Haight-Ashbury Free Medical Clinic and the Student Association for the Study of Hallucinogens (STASH) published a review of the then-current situation regarding amphetamines and related drugs in its house magazine – the *Journal of*

Psychedelic Drugs. In this they described a typical 'speed freak' as someone who had tried oral amphetamines before progressing to limited experimentation with cannabis and heroin, and finally to intravenous use of amphetamines. The first experience was almost universally described euphorically – *my whole body was just doing a physiological flip-flop.* This initial ecstatic experience led inexorably to the use of many daily injections of (perhaps) 200 mg each time, with resultant insomnia that would last for several days before a 'crash' and several days of uninterrupted sleep. On waking the whole process would be repeated. Inevitably a high degree of anorexia developed and viral hepatitis from dirty needles was not uncommon – later HIV would be a more serious complication. Other complications included a crawling sensation under the skin – *you start to feel like there's bugs going around under your skin and you know they're not there, but you pick at them anyway.* An increasing level of paranoia was also common and suicidal tendencies often developed.

The popularity of amphetamines in the San Francisco Bay area supported at this time as many as 6–8 large volume 'speed laboratories' providing perhaps 75% of the local demand, while numerous 'bathroom' operations provided the rest of the supply. These household laboratories used simple equipment available from hardware stores, with household appliances like hairdryers used to drive off the water from the initially wet products. The basic synthesis was simply a base-catalysed addition of nitromethane (or nitroethane) to benzaldehyde (or more complex aldehydes for later designer amphetamines), with reduction (using lithium aluminium hydride) of the resultant nitrostyrenes. The tricky step for the 'bathroom chemist' was, of course, the use of the lithium aluminium hydride and the ether solvent, but since the various chemicals were undoubtedly sourced from industrial or university chemistry laboratories, the 'bathroom chemists' were in reality mainly real chemists with experience of these procedures. Interestingly, the slightly impure amphetamines that arose from these operations were more highly prized than the purer versions. This impure product, known as 'flash speed', apparently produced a 'heavy flash' when used intravenously and this was characterised by a rapid and almost orgiastic pleasurable response. The purer form of the drug produced its response more slowly, though the experience was longer lasting. As with all other drugs, the initially produced drugs were often mixed or 'cut' with other inactive ingredients, *e.g.* talcum powder, baking soda, monosodium glutamate *etc.*, and much of the violence associated with amphetamine use centred around those dealers who were caught dealing in inferior products.

The drugs also began to appear in sports, and amphetamines are particularly favoured by cyclists who value the sudden and explosive boost that they provide. Sportsmen christened them *la bombe* and *la bomba*. A combination of dexamphetamine and the barbiturate amylobarbitone called Drinamyl became the mainstay of the club scene and, since this was prescribed in the form of a blue, lozenge shaped pill, they became known as 'purple hearts'. The sporadic battles that flared up in Britain between the scooter-riding 'mods' and the motorbike-riding 'rockers' in the 1960s were also claimed to have been fuelled by amphetamines. However, these beachfront spectacles were more likely due to the natural aggression of testosterone-laden males free from the strictures of national conscription, and with nothing to do on Bank Holidays, than they were to the use of drugs. In any case, most of the available drugs led to euphoria or somnolence rather than violent behaviour. Nonetheless, by the end of the 1960s, amphetamine abuse was a more serious problem in the UK than use of cocaine, heroin or cannabis. Most of the amphetamines are classified in category B of the dangerous drugs list, while methamphetamine is in class A (with heroin and cocaine).

The scene was set for two drugs that more closely resembled mescaline in structure and also in psychoactive potency. The first, MDA (3,4-methylenedioxyamphetamine) was introduced in the 1960s as an aid to psychotherapy, and the second, more infamous, drug MDMA (3,4-methylendioxy methamphetamine or 'ecstasy') first appeared on the club scene in the 1970s. This had been studied by the US chemist and pharmacologist Alexander Shulgin in the early 1960s, and he initially promoted its therapeutic benefits, but later began to write about his experiences with mescaline and the new amphetamines in a treatise entitled *PiHKAL – Phenethylamines i Have Known and Loved: A Chemical Love Story*. This was in part an autobiographical novel but it also provided full details of how to make these drugs. He also gave a comprehensive description of his experiences after taking them. After 400 mg of mescaline he reported:

Everything seemed to have a humorous interpretation. People's faces are in caricature, small cars seem to be chasing big cars, and all cars coming towards me seem to have faces . . . The effect of the drug is the extreme empathy felt for all small things . . . One cannot pluck a flower, and even to walk upon a gravel path requires one to pick his footing carefully to avoid hurting or disturbing the stones.

Not surprisingly the US Drug Enforcement Authority investigated Shulgin's laboratory and, although they could not find any evidence of criminal activity, they nonetheless labelled *PiHKAL* as a 'cookbook for the synthesis of illegal drugs'.

Ecstasy (see Figure 6.2) became especially popular in clubs and the mainstay of the 1980s rave culture (see Figure 6.3), especially in the holiday resorts of Ibiza. It induces a feeling of euphoria and empathy (even love) for others, and these feelings last for about 3–5 hours. The downside is the after-effects which include fatigue, irritability and depression, and where the user has been involved in wild prolonged and vigorous dancing, there can be serious dehydration and raised body temperature. There have been several well-publicised deaths where the ecstasy users had rehydrated to excess, leading to coma and death due to gross disruption of salt balance. However, these tragic deaths have to be seen in the context of the much larger number of deaths of young persons caused by drink-related traffic accidents.

A more dangerous drug first appeared in the Haight-Ashbury district in 1967 and this was DOM or 2,5-dimethoxy-4-methylamphetamine, one of Shulgin's 'magical half-dozen'. This produced effects akin to LSD with wild-coloured hallucinations and altered cognition, but the duration of action was about twice that of LSD, so there was more opportunity for self-harm. Mercifully this drug had a relatively short life since it was considered too dangerous for use by even hardened speed freaks.

The well-known street drug Angel Dust or PCP was first introduced as the anaesthetic phencyclidine by Parke-Davis in 1995. The drug has

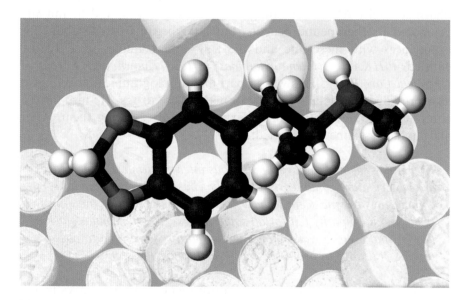

Figure 6.2   The chemical structure of ecstasy.

Figure 6.3 A 1980s rave fuelled by drugs and alcohol. © Jupiter Images.

only very limited structural similarity to the amphetamines but used to be very popular on the club scene. It is usually sprayed onto marijuana, mint or parsley leaves then smoked. In admixture with marijuana it is called Supergrass.

One very interesting amphetamine analogue is methylphenidate (Ritalin), and this shares the usual stimulant effect on the central nervous system (CNS) with the other amphetamines, but appears to possess special efficacy in the parts of the cerebral cortex that are concerned with the control of attention. The drug is thus especially effective for children with attention-deficit hyperactivity disorder (ADHD). This condition is associated with hyperactive behaviour and an inability to concentrate, and can usually be successfully treated with two to three doses of 5–20 mg of Ritalin per day.

The mode of action of all of these drugs appears to depend upon their modulation of the functions of the neurotransmitters dopamine, serotonin and noradrenaline (norepinephrine). The potentiation of dopamine function is particularly relevant to the states of euphoria attained with mescaline and the amphetamines, since dopamine is intimately involved in triggering reward responses in various parts of the brain.

None of these drugs comes close to achieving what Aldous Huxley envisaged for his drug soma in *Brave New World*, which allowed people to *take a holiday from the reality of their daily lives.* In *The Doors of Perception* he summarised what was required:

> *A new drug which will relieve and console our suffering species without doing more harm in the long run than it does in the short. Such a drug must be potent in minute doses and synthesizable . . . It must be less toxic than opium and cocaine, less likely to produce undesirable social consequences than alcohol or barbiturates, less inimical to the heart and lungs than the tars and nicotine of cigarettes.*

This wonder drug still awaits discovery, but in the meantime we can view a long history of intoxication ranging from the *peiotl* of the Aztecs to the crystal meth of Haight-Ashbury and the ecstasy of the modern club scene. Somewhere along the line, the romance of communing with the Gods has been lost. This 'romance' was eloquently described by one of the shamans who talked to Richard Evans Schultes:

> *Speak to the peyote with your hearts, with your thoughts, and the peyote sees your heart. And if you have luck, you will hear things and receive things that are invisible to others, but that God has given you to pursue your path.*

CHAPTER 7

Fly Agaric

*They are subject to various visions, terrifying or felicitous ...
owing to which some jump, some dance, others cry and suffer
great terrors ... But this applies only to those who overindulge,
while those who use a small quantity experience a feeling of extraordinary lightness, joy, courage, and a sense of energetic well-
being.*

This translation of a report from the Russian Stephan Krasheninnikov,
who travelled through Siberia in the early 1750s, was made by the
ethnobotanist R. Gordon Wasson (who worked with Albert Hofmann
on the psychedelics of *Psilocybe* mushrooms). He included it in a book
he wrote with his wife Valentina entitled *Mushrooms, Russia and History*
(1957). The cause of the intoxication was the red and white spotted
mushroom *Amanita muscaria* or fly agaric (named for its supposed
insecticidal properties, see Figure 7.1), and this held a fascination for
Gordon Wasson for much of his life. He was a strong advocate of the
theory that the 'divine soma' of ancient India was none other than fly
agaric and his later book *Soma: the Divine Mushroom of Immortality*
(1968) is devoted to this hypothesis. Soma was the name for the divine
drink taken by the deities described in the Rig Vedas – the 1000 or so
sacred hymns written in Sanskrit around 800 BC and which formed the
basis for the Hindu religion. Despite a huge amount of scholarly work
by Wasson and many other ethnobotanists and anthropologists, the

Turn On and Tune In: Psychedelics, Narcotics and Euphoriants
By John Mann
© John Mann 2009
Published by the Royal Society of Chemistry, www.rsc.org

Figure 7.1   Fly agaric. © Jupiter Images.

identity of the plant from which Soma was obtained remains in doubt. However, it is generally agreed that it was a mountain plant which had the colour of fire (*hari*), and this would probably rule out fly agaric.

What is much better understood is the use of fly agaric by the tribes of Northern Russia especially Siberia. Here we have well-substantiated reports from explorers and other visitors who all describe a beverage made from this mushroom. Probably the most cited report was that of the Swedish cartographer colonel Filip von Strahlenberg. He spent 12 years in captivity with the Koryak tribe, who were herders of reindeer in the Kamchatka peninsular of north-eastern Siberia. On his return in 1730, he related that the richer members of the tribe traded furs for quantities of fly agaric which they valued highly, and which they used on feast days:

> *When they make a feast, they pour water upon some of the mush-rooms, and boil them. Then they drink the liquor, which intoxicates them. The poorer sort . . . post themselves round the huts of the rich, and watch for the opportunity of the guests coming down to make water; and then hold a wooden bowl to receive the urine, which they drink off greedily, as having still some virtue of the mushroom in it, and by this way they also get drunk.*

Several decades later, George Steller, a German botanist and zoologist who was the naturalist on a number of expeditions to the Kamchatka peninsular between 1734 and 1741, recorded in his journals a similar account of his experiences with the Koryak. These were published in 1774 some years after his death.

> *The fly agarics are dried, then eaten in large pieces without chewing them . . . After about half an hour the person becomes completely intoxicated and experiences extraordinary visions. Those who cannot afford the high price (of the mushrooms) drink the urine of those who have eaten it, and become as intoxicated, if not more so. The urine seems to be more powerful than the mushroom, and its effect may last through the fourth or fifth man.*

Louis Lewin in his *Phantastica* confirmed these observations:

> *As soon as the Koryak notices his inebriety decreasing, he drinks his own urine . . . in this way the action may be renewed several times.*

The pharmacology of this urine drinking is obviously fascinating yet has not been properly investigated. For a long time it was believed that the main psychoactive constituent of fly agaric was muscarine, which is an agonist at one type of the receptors for the neurotransmitter acetylcholine. However, muscarine is only present in trace amounts in the mushroom. It was not until 1964–1965 that researchers in England, Japan and Switzerland conclusively proved that the major biologically active constituents were in fact ibotenic acid and its decarboxylation product muscimol, which is 5–10 times more potent. Muscimol rapidly crosses the blood-brain barrier and exerts its effects through interference with the activities of the main inhibitory neurotransmitter GABA (gamma aminobutanoic acid) – it is a competitive agonist at GABA receptors. While muscimol is rapidly metabolised – only about 30% leaves the body unchanged – ibotenic is excreted unchanged in the urine within about 90 minutes. The hallucinogenic effects of urine drinking are thus probably due primarily to the effects of ibotenic acid being converted into muscimol.

Since there are still occasional instances of mushroom poisoning with fly agaric, the clinical features of consumption are well known and can be compared with the descriptions given above. Patients typically walk as if intoxicated with alcohol and, in addition, exhibit muscle twitching and complain of hallucinations. One of the problems with fly agaric is that both patients and those using the mushroom for recreational purposes report widely differing effects. Both of the Wassons experimented with the mushroom in 1965 and 1966 and reported being disappointed with the results, especially in comparison with their experiences with *Psilocybe* mushrooms. At worst they felt nauseous and actually vomited, and at best fell into a deep sleep with vivid dreams. Other experienced psychedelic drug users have also reported disappointing results, and the use of fly agaric has largely been abandoned, or only used when alternative 'tripping' materials are not available. And in a world gone mad with legislation about health and safety, the red and white spotted mushroom has all but disappeared from children's books.

Fly agaric does not often occur in adult fiction, though H. G. Wells in his short story *The Purple Pileus* has the central character (Mr Coombes) undergo a massive character change after eating some of this eponymous mushroom. This dull and hen-pecked husband returns from a walk in the country during which he has attempted suicide using the mushroom. He then terrifies the guests at his wife's tea party, including trying to force-feed fly agaric to one of them, and then becomes the dominant partner in the marriage. Of more interest is the *soma* of Aldous Huxley's novel *Brave New World* since this was a universal panacea and hallucinogen that alleviated the strains of everyday life. However, this was a synthetic drug, and there was never any suggestion that this was a naturally occurring substance.

In *Alice in Wonderland*, Alice often eats or drinks things that make her grow or shrink. The famous caterpillar with the hookah persuades her to eat some of the mushroom on which he is seated, and after nibbling a portion she suffers another change of shape:

> *"Come, my head's free at last!" said Alice in a tone of delight, which changed into alarm in another moment, when she found that her shoulders were nowhere to be found; all she could see, when she looked down, was an immense length of neck, which seemed to rise like a stalk out of a sea of green leaves that lay far below her.*

Although both Lewis Carroll and his illustrator John Tenniel would have been familiar with Cooke's *Manual of British Fungi*, and would have known of the hallucinogenic properties of fly agaric, there is no suggestion from Tenniel's illustration that either author or illustrator had fly agaric in mind.

One piece of nonsense that has been promulgated, often accompanied by serious anthropological arguments, is that the myth of Father Christmas, with his red and white coat, is somehow associated with the use of fly agaric. The only half-serious side to this suggestion is his use of flying reindeer, since in the wilds of Scandinavia reindeer are known occasionally to consume fly agaric and thus become inebriated!

CHAPTER 8
Absinthe

There are numerous references to the use of wormwood or *Artemisia absinthium* as a vermifuge, perhaps going back to biblical times, and certainly the Elizabethan herbalist Gerard cited its use for this purpose: *wormewood voideth away the wormes of the guts.* The alcoholic brew of Tudor England called *purl*, which comprised a mixture of beer and wormwood, may also have had this activity. But here we are more concerned with the much stronger liquor whose effects Oscar Wilde summarised in his usual apposite fashion:

After the first glass you see things as you wish they were. After the second, you see things as they are not. Finally, you see things as they really are, and that is the most horrible thing in the world.

The origins of absinthe are obscure, but it is produced by distillation of a brandy-strength liquor in which the plants *Artemisia absinthium*, anise (*Pimpinella anisum*), hyssop (*Hyssopus officinalis*), fennel (*Foeniculum vulgare*) and other herbs have been steeped. The invention of this strong, yellow liquor is usually attributed to two unmarried sisters called Henriod. They lived in the Swiss village of Neuchâtel in the late 1760s and advertised their liquor as *bon extrait d'absinthe*. A little later (around 1792) a French doctor in the village of Couvet, Switzerland, by the name of Pierre Ordinaire also advertised an apparently similar liquor he called *la Fée Verte* (green fairy). What is more certain is that Major Henri Dubied purchased the recipe for absinthe from the Henriod sisters

Turn On and Tune In: Psychedelics, Narcotics and Euphoriants
By John Mann
© John Mann 2009
Published by the Royal Society of Chemistry, www.rsc.org

in 1797. He, it is claimed, valued the *bon extrait* as a general cure-all and also as an aid to his sexual potency. His daughter Emilie subsequently married Henri-Louis Pernod and the association of the names Pernod and absinthe would forever be linked. Father and son-in-law established a small distillery in Couvet, followed in 1805 by a much larger operation across the border in France, on the banks of the River Doubs in Pontarlier, which initially had an output of 400 litres of absinthe per day. With these two operations, the family business went from strength to strength throughout the nineteenth century under the commercial name of Pernod fils, though the banking family Veil-Picard gradually took over the operation during the 1890s. By 1900 the company was producing 30,000 litres every day in a well-lit and ventilated factory, with exports all over the world including the USA where absinthe was especially popular in New Orleans, San Francisco, Chicago and New York.

So how did this popularity arise? As had happened with hashish, it was the French army overseas that acquired the absinthe habit. During the various insurrections in North Africa, and especially while on duty in Algeria, the army was issued with absinthe as a vermifuge and the soldiers subsequently took their new habit back with them to France. Wormwood plants were grown primarily in the valley of the River Doubs, but also in the Jura. The anise came from as far away as the Tarn region and the fennel from the Gard. The macerated plants were initially steeped in brandy prior to distillation, then the colour was adjusted through the addition of extracts of wormwood and hyssop to produce what was described as a liquor *vert olive transparent*. The absinthe of the nineteenth century was a very strong alcoholic product (120–150 proof 60–75% alcohol by volume). Preparation of an absinthe was quite a work of art, and involved slow addition of ice-cold water through a perforated spoon bearing a lump of sugar into a glass containing about 30–40 ml of the liquor (see Figure 8.1). The resultant greenish, opalescent solution was not only the drink of soldiers and peasants, but also the favoured drink of the artists and literati who placed it centrally in their paintings or wrote of its virtues and dangers.

Of this group, perhaps the most infamous was Paul Marie Verlaine, and it was he more than anyone else who inspired the cult surrounding absinthe. Born in 1844, he was a wild youth who began drinking absinthe to excess in his early teens, and had numerous early sexual encounters with both prostitutes and other men. Despite these excesses, he published his first book of poems *Poèms Saturniens* in 1866. Soon after he became infatuated with the 17-year-old Mathilde Mauté, and married her in 1870, but their marital relations were disturbed two years

Figure 8.1 Absinthe preparation. © www.AbsintheBuyersGuide.com.

later after he received letters and poems from the young poet Arthur Rimbaud. Verlaine immediately invited Rimbaud to Paris with the words *come, dear great soul, we summon you, we await you.*

Rimbaud was born in 1854 to a mother who forced on him a serious study of the bible and Latin poetry from an early age. Perhaps as a result of this, he became an extremely gifted pupil at school and won numerous prizes before he too went off the rails as a teenager, turning to drink and petty crime. Nonetheless, he produced most of his best poetry between the ages of 16 and 21. He was already something of a rebel by the time he met Verlaine (10 years his senior), and they began a tempestuous two-year affair fuelled by absinthe and hashish. During this period Verlaine began abusing Mathilde, including (it is claimed) setting fire to her hair and clothes on one occasion, and submitting her to savage beatings. Rimbaud seems to have encouraged this abuse declaring that to be a good family man was absurd. In 1872, Rimbaud and Verlaine travelled to England but their stay was brief and impecunious, and, tiring of their relationship, Rimbaud initiated a break-up in 1873, causing a devastated Verlaine to leave for Brussels. He then contacted Rimbaud to announce that he was about to commit suicide, and Rimbaud immediately followed him to Brussels to plead with him. They met but Verlaine, completely out of his mind with drink, shot his ex-lover in the wrist. This was all too much for the local police, and in light of Verlaine's well-known reputation as a debaucher and drunk, he was sentenced to five years in jail. On his release two years later (he received remission for good behaviour) he worked for a while as a school teacher in England, and appeared to have given up his alcoholic excesses. However, after he returned to Paris in 1877, he resumed his debauched lifestyle but, in spite of this, he became something of a celebrity in the cafés and bistros of Paris, celebrated as one of the finest *fin de siècle* French poets – the so-called 'decadents' – and even received the title 'Prince of Poets' in 1894. He died aged 51 in 1896, his end hastened by a combination of cirrhosis of the liver and venereal disease. Somewhat surprisingly, although a lifelong user of his *humble ephemeral absinthe*, he claimed that opium and hashish were poisons.

Freed from the clutches of Verlaine, Rimbaud produced some excellent prose works during the period 1873–1875, and also a collection of poems under the title *Illuminations*, but then abandoned writing. For a time he became something of a world traveller and even participated in the slave trade and gun-running in Abyssinia, though he ultimately returned to Marseille suffering from what he believed to be arthritis in his knee. This unfortunately turned out to be a leg cancer and he died from the cancer in 1891 aged just 37 years.

Another inveterate absinthe drinker was the English poet Ernest Dowson, and his prose poem *Absinthia Taetra* describes rather well his experience of this liquor that certainly hastened his premature death.

The man let the water trickle gently into his glass, and as the green clouded, a mist fell away from his mind. Then he drank opaline.

Green changed to white, emerald to opal . . . and that obscure night of the soul, and the valley of humiliation, through which he stumbled, were forgotten. He saw blue vistas of undiscovered countries, high prospects and a quiet caressing sea. The past shed its perfume over him . . . He drank opaline.

At the end of his short life he seems to have revised his opinion of absinthe declaring:

On the whole it is a mistake to get binged on the verdant fluid. As a steady drink it is inferior to the homely scotch.

For a brief six-week period in 1897 he shared a close friendship with Oscar Wilde after the latter was released from Reading Jail. Wilde travelled to France but was ostracised by most of the artists and literati, though eventually befriended by Dowson when they were both staying near Dieppe. Wilde appears to have had a crush on Dowson, but the latter was not interested in romance, and he is even alleged to have persuaded Wilde to partake of the pleasures available in a local brothel. On his emergence Wilde is reputed to have said:

The first in these ten years, and it will be the last. It was like cold mutton. But tell it in England, for it will entirely restore my character.

Some idea of the warmth of this brief friendship can be found in Wilde's letters to Dowson, in one of which he reports:

I decided this morning to take a 'Pernod'. The result was marvellous. At 8.30 I was dead. Now I am alive, and all is perfect except for your absence.

But neither of them was destined to rekindle this friendship. Wilde is believed to have died from meningitis arising from an ear abscess in November 1900, shortly after the death of his friend in February of the same year aged 32 years.

In addition to these poets, it was the artists of the late-nineteenth century who have immortalised absinthe in their paintings. Probably the most famous of these is *The Absinthe Drinker* by Edouard Manet, which was presented for exhibition in the Paris Salon in 1859. However, it was rejected on the grounds that it depicted an unsavoury character in the shape of an impoverished sifter of city rubbish – an inebriated ragpicker.

It would perhaps have been more acceptable to have a professional model dressed as a ragpicker. Certainly Manet mixed with disreputable characters being a friend of Charles Baudelaire (of opium fame) from about 1858, and was probably influenced by Baudelaire's alcohol and drug-saturated lifestyle. The latter famously claimed:

One must be drunk always . . . If you would not feel the burden of time that breaks your shoulders and bows you to the earth, you must intoxicate yourself increasingly.

Manet made several sketches of Baudelaire and some have compared these likenesses with that of the famous ragpicker.

Edgar Degas also produced a painting that attracted much controversy and also distaste. In his famous *L'Absinthe* (see Figure 8.2) he depicted the actress Ellen Andrée, who apparently rarely touched alcohol, and also the engraver Marcellin Desboutin, who was also very respectable. The picture, painted in 1876, does depict brilliantly the *l'heure d'absinthe* when everything stopped for what Baudelaire termed: *the liquor that contains neither vitality nor death, neither excitation nor extinction.* The painting was severely criticised when first exhibited in Paris under the title *Au Café*. In the same year it appeared in an exhibition in Brighton, and one of the local newspaper critics described the two subjects as *a brutal sensual-looking French workman and a sickly-looking 'grisette'* (a shop girl of easy virtue). Later when the picture was exhibited (now under the title *L'Absinthe*) in London in 1893 as part of a display of contemporary artists, it attracted particular opprobrium from some critics who believed that it was an advertisement for temperance while others saw it as yet another example of French decadence.

This French decadence and the excitement of the Montmartre district of Paris were probably best depicted by Toulouse-Lautrec. His contemporary, the painter Gustave Moreau, said that *his paintings were entirely painted in absinthe*, and some of his depictions of the brothels and dance halls of Paris do have an hallucinatory glow. Absinthe appeared in many of his paintings and he was himself an inveterate absinthist and general alcoholic whose early death in 1901 (aged 36) was not unexpected. He left behind a unique record of the bars, brothels and theatres of Paris in the late nineteenth century, and probably also introduced that other hallucinatory painter Vincent van Gogh to the pleasures of brothels and absinthe. Numerous theories have been considered for van Gogh's preference for yellow in his paintings and their overall psychedelic appearance, and these range from the credible effects of glaucoma to his suspected mania, but an over-indulgence in

Figure 8.2 The painting *L'Absinthe* by Degas first exhibited in 1876 and depicting the actress Ellen Andrée and the engraver Marcellin Desboutin. © The Gallery Collection/Corbis.

absinthe may also have been contributory. Fearing that his decadent lifestyle in Paris was affecting his artistic output, van Gogh fled to Arles in 1888 (apparently on Lautrec's advice) and subsequently produced some of his best work. Paul Gaugin arrived in the summer of that year and for a while shared a house with van Gogh. They both habituated the Café de l'Alcazar which was a popular haunt of alcoholics and degenerates, and van Gogh's absinthe drinking seems to have increased. On Christmas Eve 1888, in an event reminiscent of what passed between Verlaine and Rimbaud, he is said to have feigned an attack on Gaugin with a razor before returning to their house and cutting off his ear lobe with the same razor though there has been a recent suggestion that Gaugin inflicted the wound. His behaviour led to several incarcerations in the local asylum, though he made apparent recoveries and painted some more masterpieces. His eventual decline commenced when he added turpentine consumption to his daily round of absinthe drinking, and though he was treated in the asylum at Saint Rémy de Provence from May 1889 to July 1890, his mental state deteriorated. He shot himself in the chest on 27th July and died two days later aged 37. While it is generally accepted that there was a family history of mental illness and that van Gogh was suffering from syphilis at the time of his death, there is little doubt that this was another premature death that has to be blamed at least partly on absinthe.

In contrast to these other artists, Pablo Picasso is not known to have been a great drinker of absinthe, but he did produce a number of paintings and a sculpture (*The Glass of Absinthe*) in 1914, with absinthe as their subject. He, of course, lived a long and very varied life providing good evidence that absinthe was best left as the subject of artistic output and not consumed. In his early years he had been somewhat influenced by the drinking habits of the playwright and novelist Alfred Jarry, who in turn had modelled himself on Rimbaud. Jarry came to Paris in 1891 when he was 17 and set about ruining his health with drink. He was a lifelong misogynist and probably homophilic rather than homosexual, and he apparently drank everything. His supposed only lady friend, the novelist Rachilde (a pen name) wrote that Jarry's daily consumption included as much as two litres of white wine with breakfast followed by two to three absinthes (he took it neat!) mid-morning, with more wine and absinthe with lunch, and a couple of bottles of wine with his dinner. In his writing he gives the impression that he believed absinthe had the power to enable introspection. He wrote a number of plays of which the most famous was *Ubi roi*, which was part farce and part historical drama, and he died aged 34 from the effects of TB and alcoholism. In fact the fate of most of these literary and artistic characters was determined by a combination of their unhealthy lifestyles coupled with their alcoholic intake, and in most instances a large proportion of this alcohol was taken in the form of the *green fairy*.

Fiction writers also described the use of absinthe in their books and Emile Zola's Nana is one of best known fictional characters who enjoyed all of the excesses of Montmartre. She and her friend Satin would confide in one another about their conquests and other experiences and *sometimes on afternoons when they were both in the dumps, they would treat themselves to an absinthe to help them forget the beastliness of men.*

The detrimental effects to health of excessive absinthe consumption soon attracted the attentions of the medical fraternity and a Dr Motet was one of the first to describe the pallid, weak and enfeebled creatures who populated the Parisian Psychiatric Hospital Bicêtre. His report entitled *Considérations générals sur l'alcoolisme et plus particulièrement des effets toxiques produits sur l'homme par la liqueur d'absinthe* appeared in 1859. But it was Dr Valentin Magnan who claimed in 1869 to have demonstrated the toxic effects of wormwood, and by implication the major constituent thujone. In the description of his experiment he reported that a guinea pig in a glass cage with a saucer of wormwood extract, and another with a saucer of alcohol, were left to inhale the vapours. The former after initial excitement suffered convulsions while the latter became intoxicated. He also described similar experiments with a cat and a rabbit. Dogs were given large oral doses of absinthe and then suffered epileptic-type seizures, while alcohol simply caused intoxication. This led him to propose that thujone was the culprit, though he did not exclude the possibility that constituents of hyssop and anise might be contributory. The absinthe industry responded by producing the liquor *sans thujone*, but these apparent risks seem to have provided little discouragement for the populace since by 1910 the French were consuming more than 35 million litres each year.

Increasingly, governments became alarmed at the rise of alcoholism, especially involving absinthe drinking, and its association with criminal behaviour. In Switzerland absinthe was banned in 1910 following a well-publicised murder case involving a vineyard worker called Jean Lanfray. By all accounts Jean Lanfray consumed alcohol in prodigious quantities, and on the day in August 1905 when he shot and killed his wife and two young daughters, he had consumed several brandies, at least a couple of bottles of wine and two large glasses of absinthe. The effects and dangers of this last drink attracted most of the press coverage during the trial in 1906. Some of the lawyers and journalists supported the manufacturers in their claims that absinthe was being unduly blamed for the actions of a very obvious general alcoholic but, despite this support, a move to pass the laws banning absinthe production moved inexorably through the Swiss parliamentary system.

The temperance movement in France became an increasingly dynamic force and it published statistics showing that the number of people in asylums had increased from around 50,000 in the 1870s to more than 100,000 in the early 1900s, with the majority of these having alcohol (absinthe)-related problems. In 1888 the French *Ligue National Contre l'Alcoolisme* obtained 400,000 signatures for a petition declaring that absinthe provoked criminal activities and epilepsy and was ruinous for family life. They also produced some verse that supported their claims:

Je suis la Fée Verte,
Ma robe est le couleur d'espérance,
Je suis la ruine et la doleur,
Je suis la honte,
Je suis le déshonneur,
Je suis la mort,
Je suis l'Absinthe.

Debate raged in the French National Assembly for many years and not surprisingly the absinthe producers mounted a vigorous lobby stressing the value to the exchequer of the taxes raised through the sale of absinthe. These exceeded 50 million francs a year by 1910, and this argument was successful right up until the start of the First World War. Now the generals raised objections to the sale of absinthe claiming that they needed sober troops to fight the enemy, and the French Government was forced to enact a law in March 1915 to close down the factories, with essentially no compensation paid to the manufacturers. The tide had already turned against absinthe, partly because of a spate of grisly murders perpetrated by alleged *absinthists*, but also because there were growing fears about the declining birth rate in France which was blamed upon the supposed sterilising effects of absinthe.

France was the last major country to ban absinthe, since the USA introduced a ban in 1912. There was no reason to have a ban in the UK because absinthe drinking had not been a problem. Spain also allowed production of absinthe to continue and Pernod moved their production to Catalonia, while maintaining anise-based liquor production in France in the form of their product *pastis*. Interestingly, since about 2000, many countries, including France, have allowed a revival of absinthe production and sale, and this followed an evaluation of the toxicological properties of thujone by a FAO/WHO commission entitled *Codex Alimentarius* in 1979. This commission recommended that thujone could be used as a flavouring (bittering) material, provided that its concentration did not exceed 10 mg of thujone per kg of alcoholic beverage (25% maximum volume of alcohol) and, more recently, this has been amended to 35 mg/kg. The USA has thus far refused to accept these recommendations; the maximum amount of thujone allowed in any alcoholic beverage is 10 mg/litre of beverage. The upshot of these recommendations is the appearance of a plethora of new products with names like

Moulin Rooz and la Fée Absinthe – names that are calculated to revive and extend the cult status of the liquor.

So can we separate myth from reality or the genuine alcoholic from the more ephemeral absinthist? The dogma originally established by Dr Magnan was that thujone was the toxic ingredient and led to the state of absinthism. In his earliest publication in 1864 for *l'Union Médicale* under the title *Accidents déterminés par la liqueur d'absinthe* he reported a study of a patient who regularly drank five to six glasses of absinthe each day together with quantities of wine and *eau-de-vie*. This man entered the hospital on several occasions with what Magnan described as *delirium tremens* though seemed to progress to a state where he suffered extreme hallucinations and epileptic seizures. Magnan summarised his case and blamed these later manifestations on absinthe:

Il commence par des excès de vin et d'eau-de-vie et devient alcoholique; puis, il s'adonne à absinthe et devient épileptique.

As mentioned earlier, his comparative experiments on animals seemed to confirm that absinthe rather than alcohol *per se* could induce epileptic seizures, though he did administer gram quantities of the liquor either mixed in with bread or directly into the veins of the animals. There was considerable controversy over the next 25 years, and contradictory results were obtained by two very brave physiologists (Cadéac and Meunier) from Lyon who self-administered a gram of what was described as *essence of absinthe* and reported a stimulant effect. They calculated that they had consumed the equivalent of 60–200 glasses of absinthe in this experiment. Not withstanding these and other apparently contradictory findings, Magnan waged an incessant war against alcoholism and what he believed to be the associated condition of absinthism.

Certainly thujone is known to possess activity in the brain and can lead to epileptic-like seizures at a dose of several hundred milligrams, and over the years various claims have been made for thujone levels in absinthe as high as 260 mg/litre, which would mean around 9 mg of thujone in a typical 35 ml glass of neat absinthe. However, a definitive and fascinating study was carried out by the German group of Lachenmeier in 2007 (reported in 2008) on 13 bottles of pre-ban (*i.e.* pre-1915) absinthe which revealed typical thujone contents of 0.5 to 48 mg/litre with a mean value of 33 mg/litre. These vintage bottles of absinthe were sourced from as far back as 1895 from Edouard Pernod of Switzerland and Pernod Fils of France. They also analysed some post-ban absinthe (mainly from Spain) including some illicit absinthe from

Val-de-Travers in Switzerland, and none of these contained more than 33 mg/litre.

Two features of their analytical results are particularly noteworthy: the variability of the thujone levels and the quite large amounts of the monoterpene fenchone (up to 45 mg/litre) that were also present. Neither of these findings is surprising since the fenchone comes from hyssop, and any herbal extract will always differ in its composition depending upon the source of the plant and the time of harvesting. This is the perennial problem with herbal medicines where the consumer has no way of assessing the variability of the preparations and consequently their likely efficacy. What is clear is that there was no sign of the large amounts of thujone claimed for nineteenth-century absinthe, although the long-term harmful effects of say 9 mg of thujone per glass should not be underestimated. However, compared with the vast amounts of ethanol that were being consumed contemporaneously, the amounts of thujone pale into insignificance.

But what about the *bewilderment of the mind* and *the blue vistas of undiscovered countries* claimed by Ernest Dowson? An apparent explanation for these effects was provided by del Castillo and Anderson in a *Nature* paper in 1975, when they suggested that thujone and tetrahydrocannabinol (from cannabis) produced similar psychotropic responses. They speculated that the three-dimensional structures of the two molecules were sufficiently similar to interact with common sites in the brain. At this time the cannabinoid receptors had not been identified, but subsequent research by Meschler and Howlett in 1999 showed convincingly that thujone did not bind to CB1 receptors in rat brain. Furthermore, rats given thujone behaved differently from rats given the synthetic cannabinoid levonantradol.

Of course this all rather spoils the story since it is likely that Dowson, Rimbaud, Verlaine and all the other absinthists were in reality nothing more than hardened alcoholics, and that the exciting mental effects claimed for *la Fée Verte* were little more than those associated with intoxication. Not that these revelations and consequent reality check will do anything to dampen the enthusiasm for the cult of absinthe, and five minutes spent on the internet reveals a veritable bazaar of absinthe-related products and publications.

This book commenced with an account of LSD, probably the most potent hallucinogen ever encountered by human cultures, and ends with a putative euphoriant which was something of a damp squib! The one feature that these two products, and all of the other substances mentioned in this book, share is the colourful characters who have popularised their use. These people and the drugs they have used and abused have had a huge impact on history, politics, social structure and culture. The individual psychedelics, narcotics and euphoriants are all now illegal or tightly controlled, so recreational use is fraught with danger. However, the potential medical use of these substances is largely unexplored, and it is not impossible that in the future an alternative form of Timothy Leary's slogan might read *turn on, tune in and feel the (clinical) benefit*.

Further Reading

General

R. Davenport-Hines, *The Pursuit of Oblivion*, Weidenfeld and Nicolson, London, 2001.

W. Emboden, *Narcotic Plants*, Studio Vista, London, 1972.

J. G. Hardman, L. E. Limbird, P. B. Molinoff, R. W. Ruddon and A. G. Gilman (ed.), *Goodman and Gilman's The Pharmacological Basis of Therapeutics*, McGraw-Hill, New York, 9th edn, 1996.

M. Hesse, *Alkaloids: Nature's Curse or Blessing*, Wiley-VCH, Zurich, 2002.

J. Mann, *Murder, Magic and Medicine*, Oxford University Press, Oxford, 2000.

G. L. Patrick, *An Introduction to Medicinal Chemistry*, Oxford University Press, Oxford, 3rd edn, 2005.

R. E. Schultes and A. Hofmann, *Plants of the Gods*, van der Marck Edition, New York, 1979.

W. Sneader, *Drug Discovery*, Wiley, Chichester, 2005.

From Ergot to LSD

F. J. Bové, *The Story of Ergot*, Karger, Basel, 1970.

L. R. Caporael, Ergotism: the satan loosed in Salem, *Science*, 1976, **192**, 21–26.

Turn On and Tune In: Psychedelics, Narcotics and Euphoriants
By John Mann
© John Mann 2009
Published by the Royal Society of Chemistry, www.rsc.org

J. Gonzalez-Maeso *et al.*, Hallucinogens recruit specific cortical 5-HT$_{2A}$ receptor-mediated signalling pathways to affect behaviour, *Neuron*, 2007, **53**, 439–452.

A. Hofmann, *LSD: My Problem Child*, McGraw-Hill, New York, 1980.

A. Hofmann, in *Plants in the Development of Modern Medicine*, ed. T. Swain, pp. 235–260, Harvard University Press, 1972.

A. Hofmann and A. Stoll, in *The Alkaloids*, ed. R. H. F. Manske, **VIII**, pp. 726–783, 1965.

A. Hofmann, C. A. P. Ruck and R. G. Wasson, *The Road to Eleusis – Unravelling the Secret of the Mysteries*, Harcourt Brace Jovanovich, New York, 1979.

J. Higgs, *I Have America Surrounded: the Life of Timothy Leary*, Friday Books, London, 2006.

M. K. Matossian, The time of the great fear, *The Sciences*, 1983, Feb/March, 38–41.

P. W. J. van Dongen and A. N. J. de Groot, History of ergot alkaloids from ergotism to ergometrine, *Eur. J. Obstet. Gynecol.*, 1995, **60**, 109–116.

Opium

V. Berridge and G. Edwards, *Opium and the People*, Allen Lane, London, 1981.

M. Booth, *Opium: A History*, Simon and Schuster, London, 1996.

T. De Quincey, *Confessions of an English Opium Eater*, Dover Publications, New York, 1995.

J. Hughes *et al.*, Identification of two pentapeptides from the brain with potent opiate agonist activity, *Nature*, 1975, **258**, 577–579.

A. W. McCoy, *The Politics of Heroin: CIA Complicity in the Global Drugs Trade*, Lawren Hill, Chicago, 2nd edn, 2003.

P. D. Scott, *Drugs, Oil and War*, Rowman and Littlefield, Lanham, Maryland, 2003.

Cocaine

M. Bowden, *Killing Pablo*, Atlantic Books, London, 2001.

P. M. Ellis and A. T. Dronsfield, Cocaine – a short trip in time, *Education in Chemistry*, 2007, 142–146.

S. Hyde and G. Zanetti (ed.), *White Lines: Writers on Cocaine*, Thunder's Mouth Press, New York, 2002.

M. M. King, Dr. John Pemberton: originator of Coca-Cola. *Pharmacy in History*, 1987, **29**, 85–89.

R. T. Martin, Role of coca in the history, religion and medicine of South American Indians, *Econ. Bot.*, 1970, **24**, 422–428.

R. Sabbag, *Snowblind; a Brief Career in the Cocaine Trade*, Canongate, Edinburgh, 1998.

D. Streatfield, *Cocaine*, Virgin Books, London, 2002.

T. Weiner, *Legacy of Ashes: History of the CIA*, Allen Lane, London, 2007.

Cannabis

M. Booth, *Cannabis*, Bantam Books, London, 2003.

B. Costa, On the pharmacological properties of Δ^9-tetrahydro-cannabinol, *Chem. Biodiversity*, 2007, **4**, 1664–1677.

E. Di Tomaso, M. Beltramo and D. Piomelli, Brain cannabinoids in chocolate, *Nature*, 1996, **382**, 677–678.

D. Gaskell, Chocolate – melting the myth, *Chemistry in Britain*, April 1997, 32–34.

L. O. Hanus, Discovery and isolation of anandamide and other endocannabinoids, *Chem. Biodiversity*, 2007, **4**, 1828–184.

R. Mechoulam, Le Cannabis, *La Recherche*, 1976, **7**, 1018–1026.

R. Mechoulam (ed.) *Marijuana: chemistry, pharmacology, metabolism and clinical effects*, Academic Press, New York, 1976.

R. Sabbag, *Smokescreen*, Canongate Books, Edinburgh, 2003.

Belladonna, Mandrake and Daturas

A. J. Carter, Myths and mandrake, *J. Royal Soc. Med.*, 2003, **96**, 144–147.

M. Murray, *The Witch and Demonology Cult in Western Europe*, Oxford University Press, Oxford, 1921.

J. B. Russell and B. Alexander, *A New History of Witchcraft*, Thames and Hudson, London, 2007.

M. Summers, *History of Witchcraft and Demonology*, Routledge, London, 1973.

R. W. Thurston, *The Witchhunts: a History of the Witch Persecutions in Europe and N. America*, Longman, 2006.

Peyote

K. Harries-Rees, Taking a medical trip, *Chemistry World*, September 2007, 68–71.

A. Huxley, *The Doors of Perception*, Chatto and Windus, London, 1954.

L. Iverson, *Speed, Ecstasy, Ritalin: the Science of Amphetamines*, Oxford University Press, Oxford, 2006.

H. Michaux, *Misérable Miracle: la Mescaline*, Editions Gallimard, Paris, 1972.

Absinthe

B. Conrad III, *Absinthe: History in a Bottle*, Chronicle Books, San Francisco, 1988.

J. Del Castillo, M. Anderson and G. M. Rubottom, Marijuana, absinthe and the central nervous system, *Nature*, 1975, **253**, 365–366.

D. W. Lachenmeier *et al.*, Chemical composition of vintage preban absinthe, *J. Agric. Food Chem.*, 2008, **56**, 3073–3081.

J.-P. Luauté, L'Absinthism: la faute du docteur Magnan, *L'Évolution Psychiatrique*, 2007, **72**, 515–530.

J. P. Meschler and A. C. Howlett, Thujone exhibits low affinity for cannabinoid receptors, *Pharmacol. Biochem. Behav.*, 1999, **62**, 473–480.

Glossary

NEUROTRANSMITTERS

All of the substances in this book interact in some way with neuro-transmission at the point where a nerve impulse passes across a synapse (gap between two neurons) or a neuroeffector junction (gap between a neuron and another type of brain cell, *e.g.* astrocyte). Typically the first nerve impulse results in the release of a stored neurotransmitter substance which leaks across the synapse or neuroeffector junction to impinge on a protein receptor on the second cell. These neuro-transmitters are all small molecules and once they have bound to the receptor they elicit a response. These can be very complex but typically they involve the release of effector chemical substances or a change in the levels of sodium, potassium, calcium or chloride ions. Once the neurotransmitter has produced the effect, it is either destroyed or reabsorbed into the storage vesicles from which it was released. This is known as **re-uptake**.

The most important neurotransmitters in the brain that are affected by the psychotropic substances described in this book are acetylcholine and noradrenaline (norepinephrine) (also important in the rest of the central nervous system), dopamine, 5-hydroxytryptamine (5-HT or serotonin) and GABA (gamma aminobutanoic acid).

Turn On and Tune In: Psychedelics, Narcotics and Euphoriants
By John Mann
© John Mann 2009
Published by the Royal Society of Chemistry, www.rsc.org

Noradrenaline receptors are particularly prevalent in the cerebral cortex, hippocampus and cerebellum, and these areas are especially important for control of wakefulness and alertness, blood pressure regulation and control of mood.

Dopamine receptors are most prevalent in the corpus striatum, the limbic system and the hypothalamus, and the neurons in these areas are believed to be implicated in the control of motor function, emotion and reward, and the control of hormone secretion from the pituitary.

5-HT receptors have a similar distribution to those for noradrenaline and are implicated in the control of behaviour, sleep induction and the degree of wakefulness, mood, feeding behaviour and sensory responses.

Acetylcholine receptors are widely distributed throughout the brain and also all other parts of the body. In the brain, they are especially implicated in learning and short-term memory, the state of arousal and motor control. There are two sub-types of acetylcholine receptors identified because they interact with the natural ammonium species muscarine or the natural alkaloid nicotine – hence muscarinic and nicotinic receptors.

GABA receptors are common throughout the brain especially in the nigrostriatal system. GABA is the major inhibitory neurotransmitter in the brain and is the target for the benzodiazepines (anti-depressants), barbiturates and muscimol from fly agaric. All of these psychotropic substances potentiate the inhibitory functions of GABA.

THE BLOOD-BRAIN BARRIER

The brain is not easily penetrated by polar molecules, for example by most anti-cancer drugs and antibiotics. It is much more permeable to lipophilic molecules like most of the substances described in this book. A good example of the differing ease of uptake and resultant differences in levels of potency is provided by the lipophilic drug heroin (very potent) and the more polar morphine (less potent).

FROM ERGOTISM TO LSD

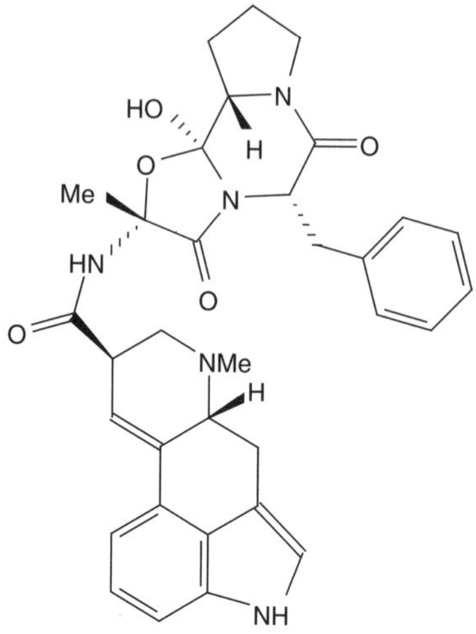

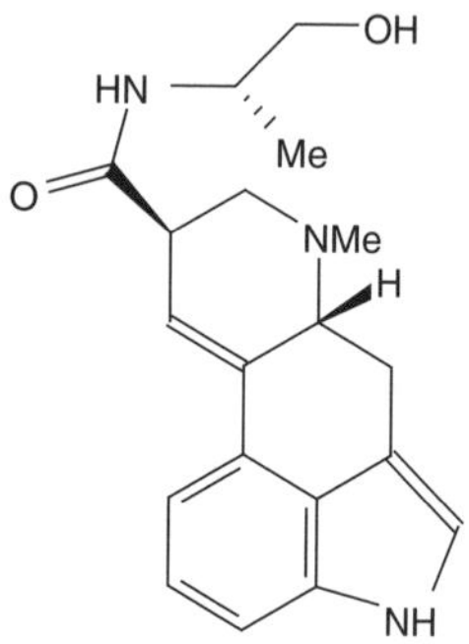

ergotamine

ergometrine

Ergotamine was isolated from *Claviceps purpurea* by Stoll in 1918, but its structure was not determined until 1951 by Stoll and Hofmann. It was synthesised by Hofmann in 1961. It has vasoconstrictive activity on blood vessels with special activity in the uterus, though it has to be used with care since doses of more than 20 mg per week can lead to symptoms of ergotism. It has also been used in the treatment of migraine and appears to act at 5-HT receptors helping to counteract changes to the blood flow that occur before and during an attack of migraine.

Ergometrine was isolated from *Claviceps purpurea* in 1935, more-or-less simultaneously by the groups of Kharasch, Dudley and Moir, Thompson, and Stoll and Burckhardt. It has been used widely in the treatment of postpartum haemorrhage due to its stimulation of smooth muscle in the uterus with resultant production of rhythmic contractions.

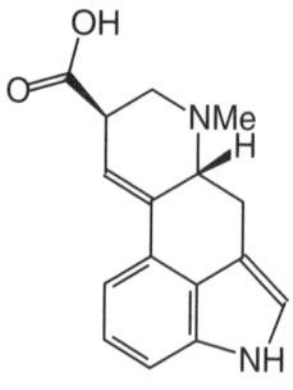

lysergic acid 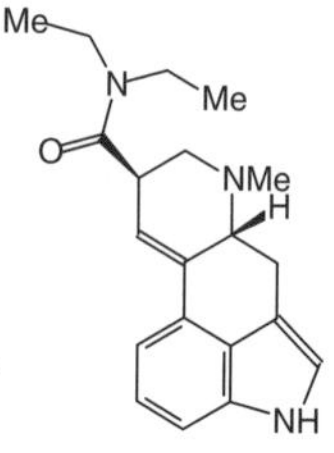   LSD 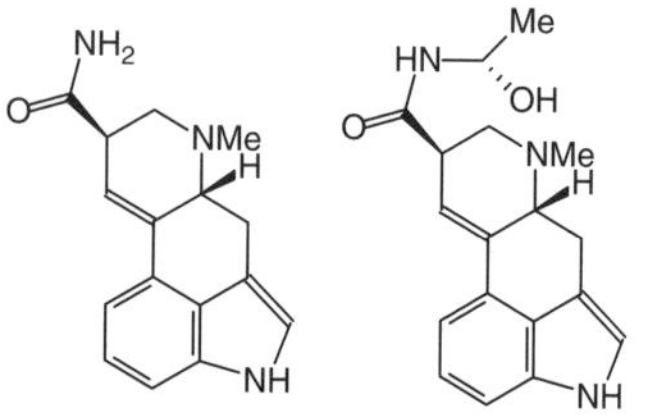   *ololiuqui* alkaloids

Lysergic acid is the parent structure of all of the ergot alkaloids and was produced in 1934 by Jacobs and Craig through the alkaline hydrolysis of ergotinine. It was synthesised by Woodward in 1954.

LSD was first produced in 1938 by Stoll and Hofmann by reaction of lysergic acid chloride with diethylamine. It is hallucinogenic at doses as low as 25–50 μg.

Lysergic acid amide and hydroxyethylamide of lysergic acid were identified by Hofmann in 1960 from *ololiuqui*. They have about one-twentieth of the psychotropic activity of LSD.

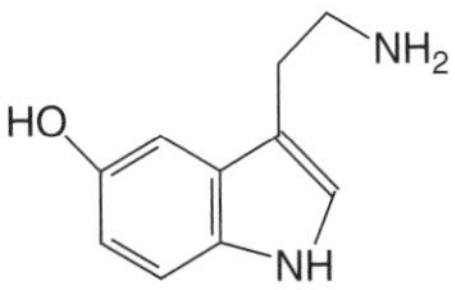

5-hydroxytryptamine
(serotonin, 5-HT)

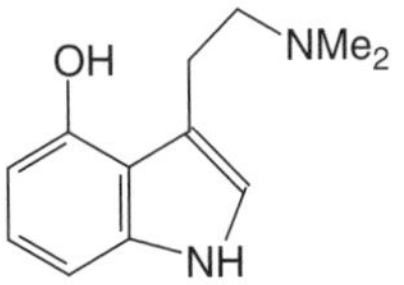

psilocin

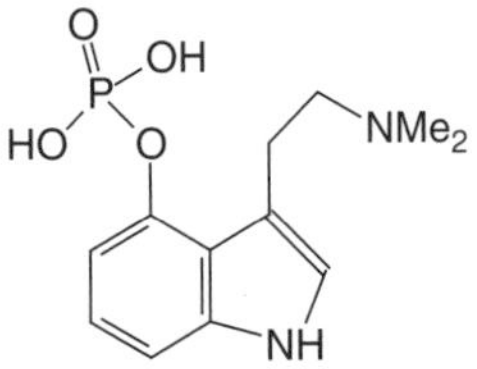

psilocybin

5-HT is a major neurotransmitter both in the brain and in peripheral nerves.

Psilocin and **psilocybin** were first isolated by Hofmann in 1958 from the magic mushroom *Psilocybe mexicana*, believed to be the basis for the magical preparation *teonanacatl*. Like LSD, they are especially active at the subclass of receptors 5-HT$_2$.

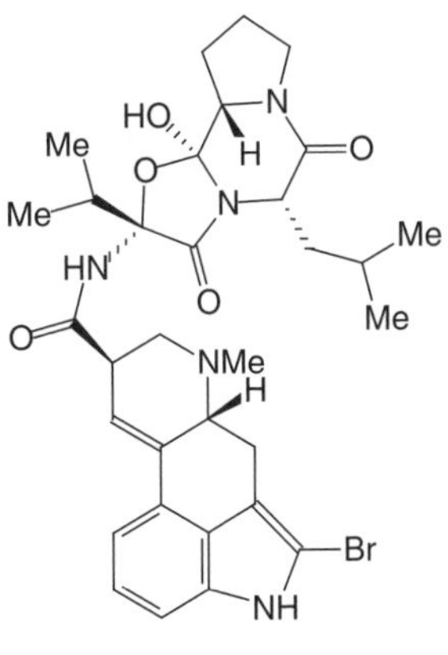

bromocryptine

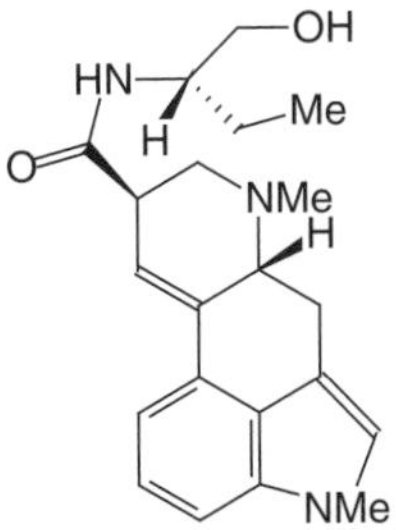

methysergide

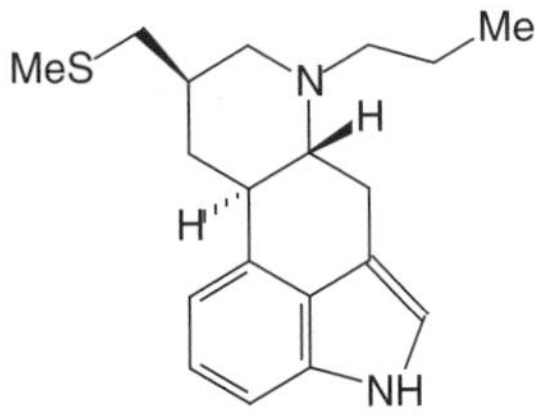

pergolide

Bromcriptine was synthesised in 1968 and has dopamine-like activity. It is orally active and is used to treat patients suffering from Parkinson's disease who have become resistant to the effects of DOPA.

Methysergide was first synthesised (by Hofmann) in 1960. It is an antagonist of 5-HT and is used in the prevention of migraine.

Pergolide was first synthesised in 1979 and also has dopamine-like activity. It is used in conjunction with DOPA for the treatment of Parkinson's disease.

OPIUM, MORPHINE AND HEROIN

morphine   codeine   thebaine   heroin

Morphine was first isolated by Sertürner in 1806, the structure was determined by Gulland and Robinson in 1925 and it was first synthesised by Gates in 1952.

Codeine and **thebaine** are the other major alkaloids from *Papaver somniferum* (the opium poppy). Codeine has around 20% of the analgesic activity of morphine.

Heroin was synthesised by acetylation of morphine by Wright in 1874. It crosses the blood-brain barrier more effectively than morphine and is then metabolised to morphine.

etorphine   buprenorphine

Etorphine and **buprenorphine** were synthesised by Bentley in 1963. The former is around 1,000 times more potent than morphine, and the latter is about 25–50 times more potent with a slower onset of respiratory depression.

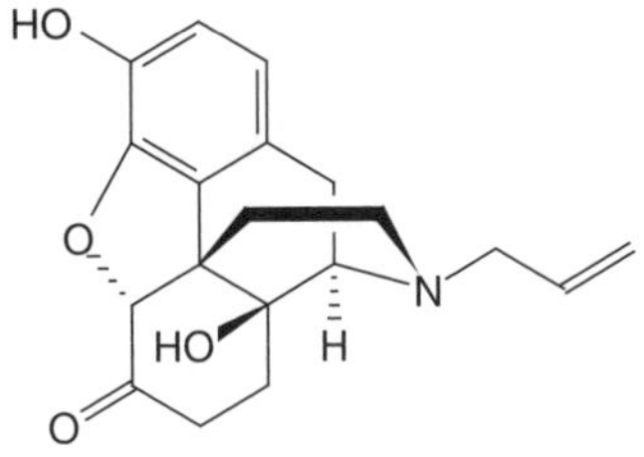

naloxone

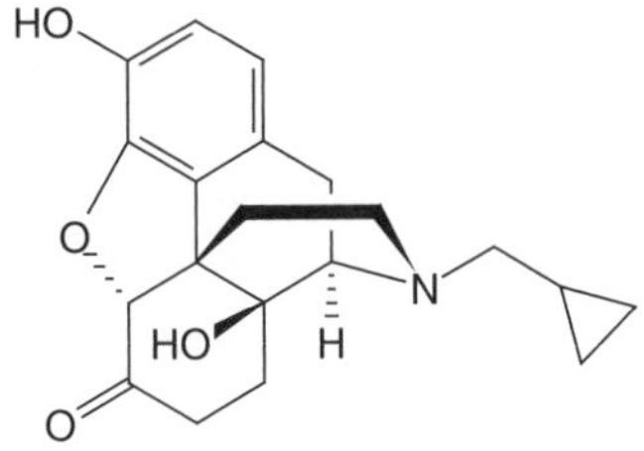

naltrexone

Naloxone was synthesised in 1961 and is a pure opiate antagonist, that is it blocks the actions of opiates and the endogenous peptides (enkephalins, endorphins *etc.*). It is used to reverse the effects of overdose and also the respiratory depression caused by morphine and heroin.

Naltrexone was synthesised in 1967 and has similar activity to naloxone, but has a much longer duration of activity (half-life in plasma of 10 hours).

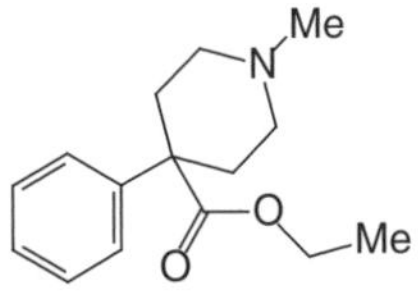

pethidine

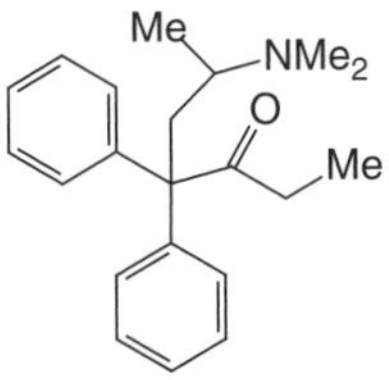

methadone

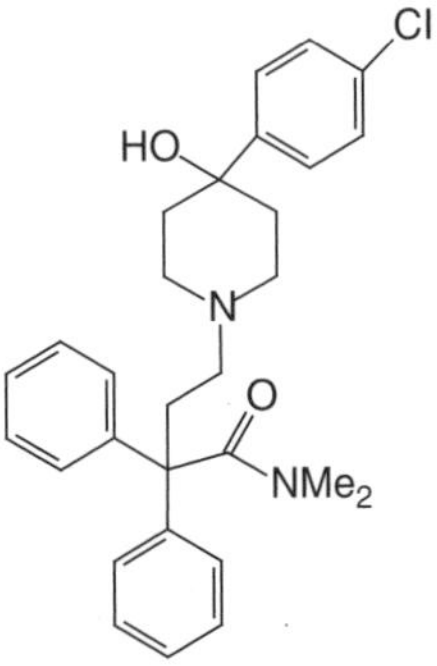

loperamide

Pethidine has analgesic activity about one-tenth that of morphine but has a very rapid onset of activity and a shorter duration of action especially in the newborn.

Methadone has similar analgesic and narcotic activity to morphine but has a longer duration of action with a plasma half life of 15–40 hours.

Loperamide (Imodium) has marked anti-diarrhoeal activity since it diminishes the peristalsis and secretion in the gut.

The **endogenous opioid peptides** include **met-enkephalin** (H_2N-Tyr-Gly-Gly-Phe-Met-CO_2H) and **leu-enkephalin** (H_2N-Tyr-Gly-Gly-Phe-Leu-CO_2H); **endomorphin-1** (H_2N-Tyr-Pro-Trp-Phe-$CONH_2$) and **endomorphin-2** (H_2N-Tyr-Pro-Phe-Phe-$CONH_2$); and the **endorphins** and **dynorphins** which are polypeptides of up to 33 amino acids.

COCA AND COCAINE

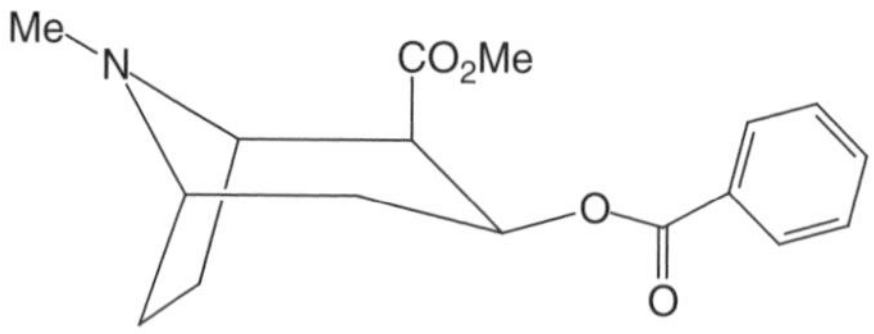

cocaine

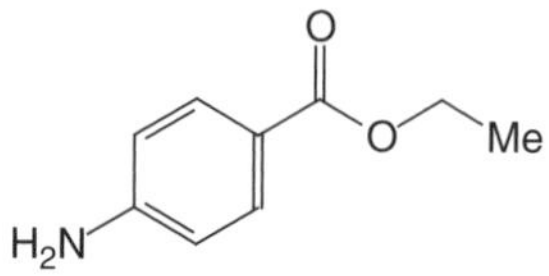

benzocaine

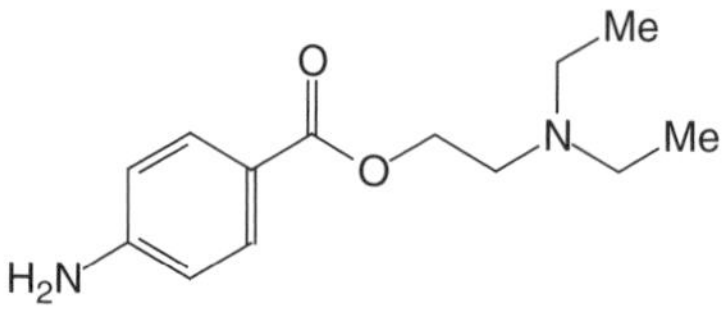

procaine

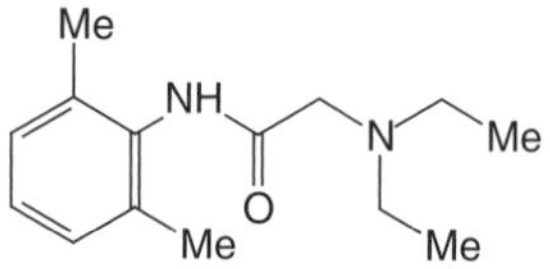

lidocaine

Cocaine was isolated by Niemann in 1860 with its structure identified and the molecule synthesised by Willstätter in 1898. It inhibits the re-uptake of dopamine and produces similar effects to the **amphetamines** but has less effect on movement (locomotor effects). The local anaesthetics which include **benzocaine** (Ritsert, 1890), **procaine** (Einhorn, 1906) and **lidocaine** (Löfgren, 1946), as well as cocaine, bind to the membranes of nerve cells and block the opening of the sodium channels following stimulation by acetylcholine or another neurotransmitter. This effectively abolishes onwards transmission of a nerve impulse.

CANNABINOIDS

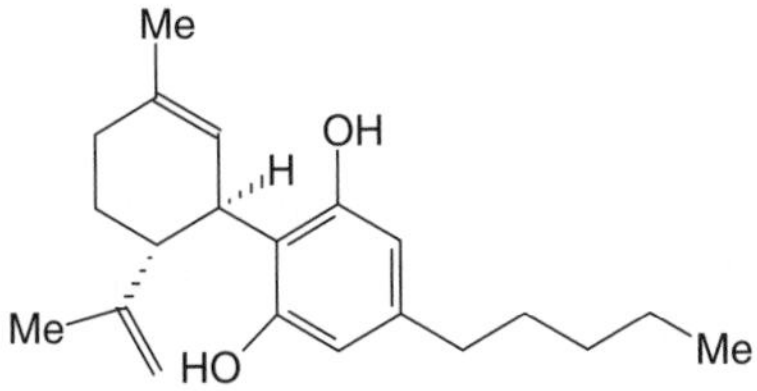

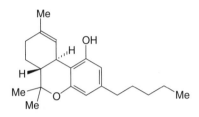

cannabidiol

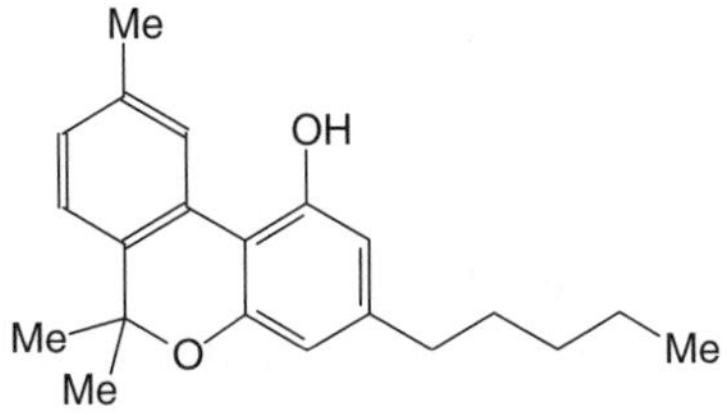

cannabinol

$\triangle^9$- tetrahydrocannabinol

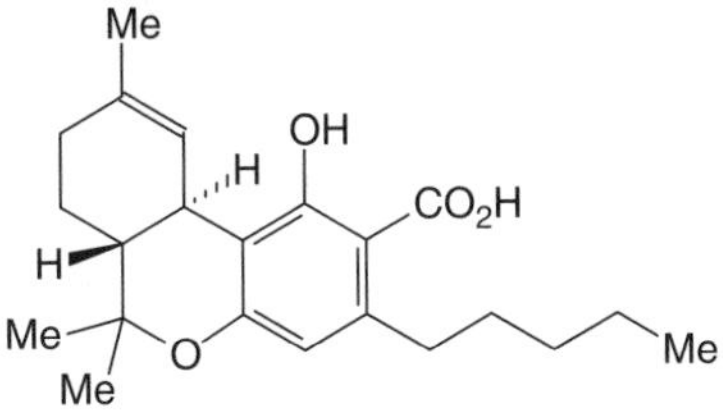

$\triangle^9$-tetrahydrocannabinolic acid

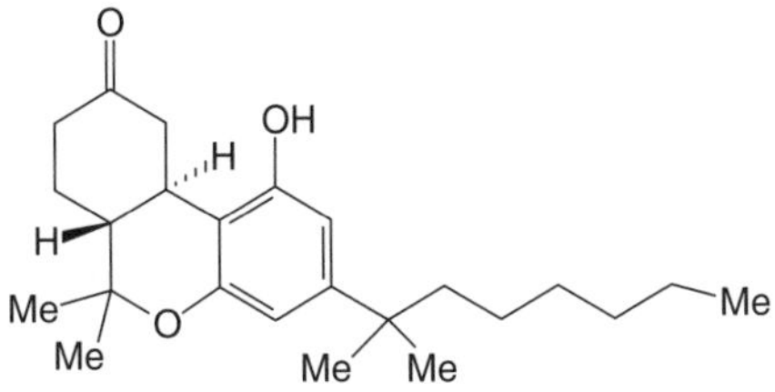

nabilone

Of the cannabinoids from *Cannabis sativa* only $\triangle^9$-**tetrahydrocannabinol** (first identified by Gaoni and Mechoulam in 1964) has significant psychotropic activity. **Nabilone** is a synthetic cannabinoid, first prepared in 1975, and is a very effective anti-emetic agent used for the treatment of drug-induced emesis in cancer and AIDS patients.

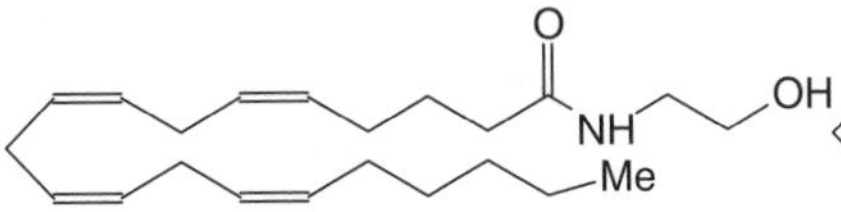

anandamide

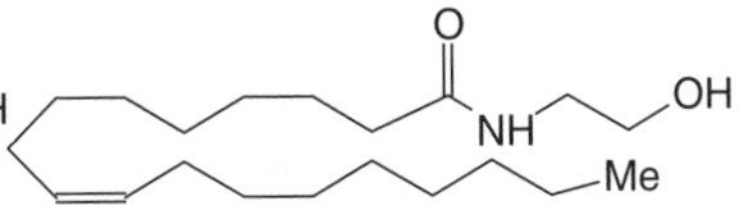

oleoylethanolamide

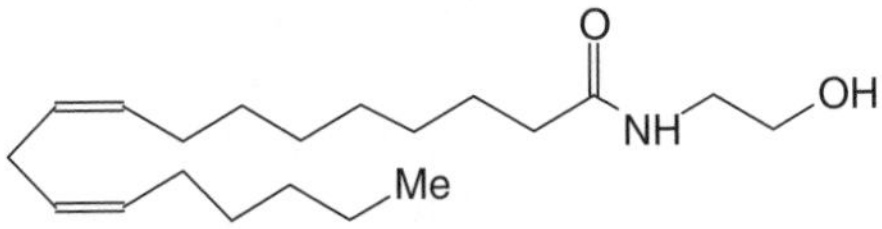

linoleoyl ethanolamide

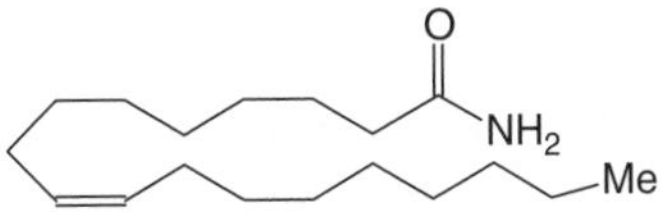

oleamide

Anandamide was the first endogenous cannabinoid to be discovered by Mechoulam in 1992.

Both anandamide and **linoleoyl ethanolamide** were reported to be present in chocolate, but this was subsequently disproved. However, **oleoyl ethanolamide** is an endogenous regulator of food intake, and **oleamide** is an endogenous regulator of sleep patterns.

BELLADONNA, MANDRAKE AND DATURAS

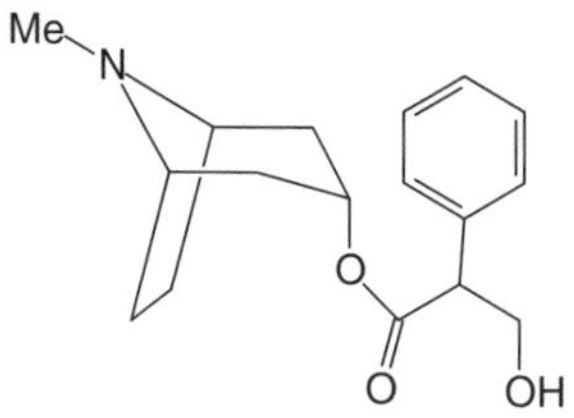

atropine

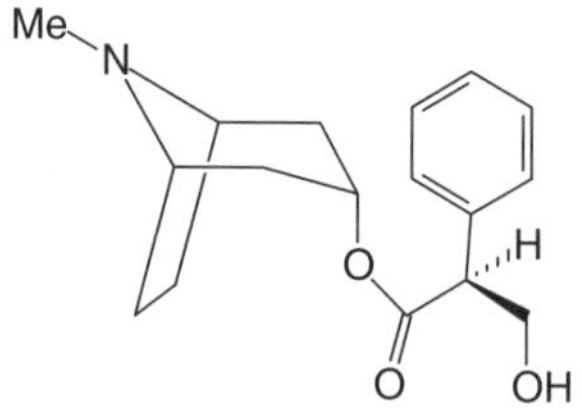

hyoscyamine

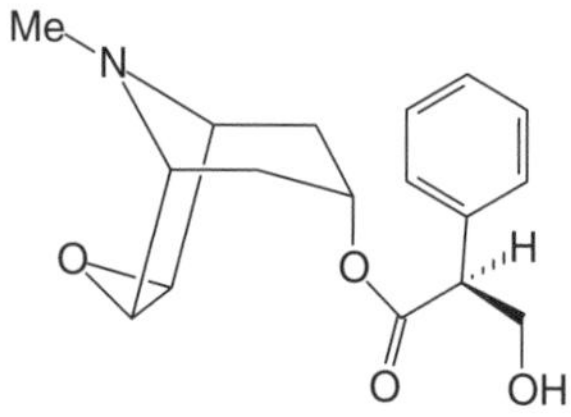

hyoscine (scopolamine)

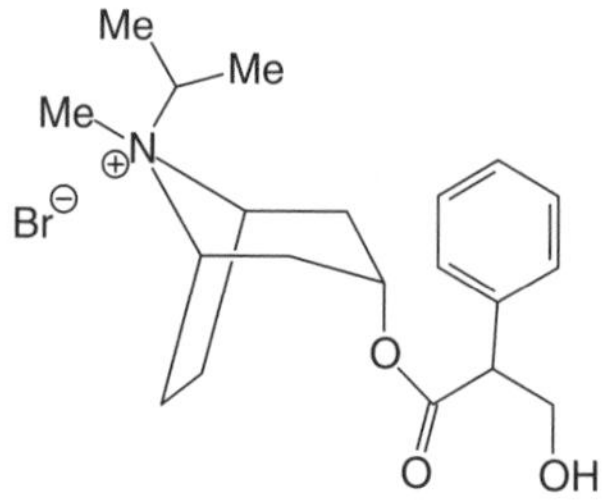

ipratropium bromide (Atrovent)

Atropine is the name used for the mixture of stereoisomers isolated from *Atropa belladonna* and other plants like *Mandragora officinarum* and *Hyoscyamus niger*. It was first isolated in 1819 by Runge; the structure was elucidated by Ladenburg in 1883; and it was synthesised for the first time by Willstätter in 1902. The pure stereoisomers present in the plants before isolation procedures are used are called **hyoscyamine** and **hyoscine**. These **tropane alkaloids** are all competitive antagonists at the muscarinic class of acetylcholine receptors. Atropine has limited activity in the central nervous system, since it is poorly taken up through the blood-brain barrier. Hyoscine is the most effective at crossing the blood-brain barrier and in consequence has the most potent activity as a euphoriant and disorientating agent. In larger doses it causes serious CNS depression, narcosis and death. Its action at muscarinic receptors in the gut provides the basis of the anti-emetic effect of scopolamine plasters (to prevent travel sickness).

Ipratropium bromide has marked bronchodilator activity and is used in asthma therapy.

PEYOTE AND AMPHETAMINES

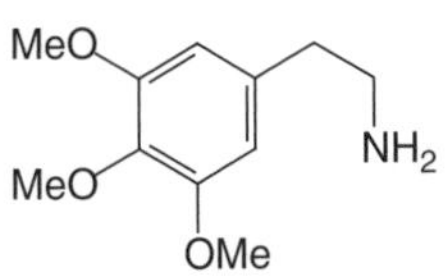

mescaline 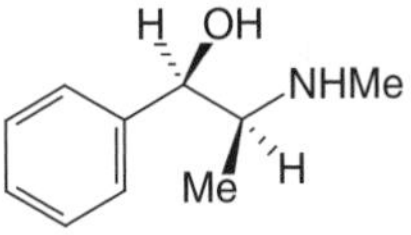   ephedrine 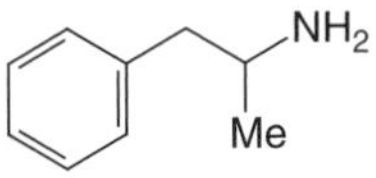   amphetamine (Benzedrine)

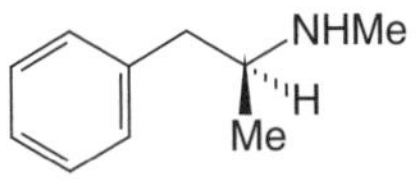

methedrine 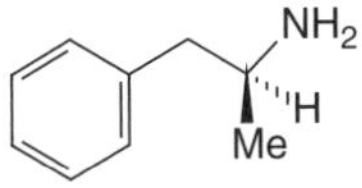   dexedrine

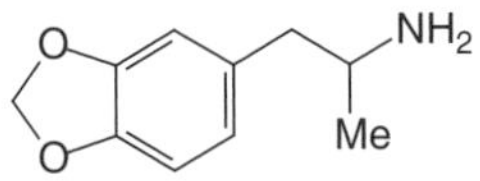

MDA 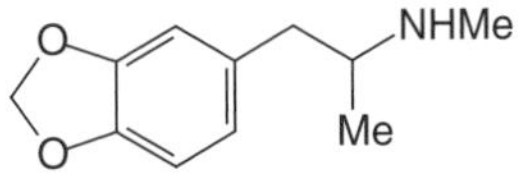   MDMA (Ectasy) 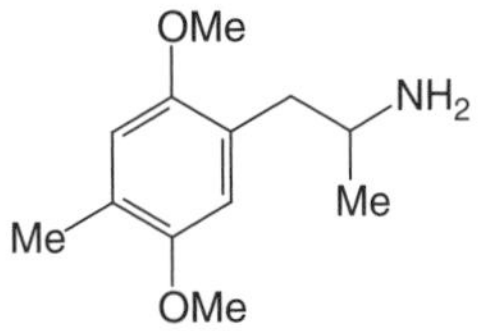   DOM

Mescaline is the major psychoactive constituent of the peyote cactus *Lophophora williamsii*. **Ephedrine** is the major biologically active constituent of *Ephedra sinaica* and its bronchodilator activity was originally of some interest. However, its stimulant activity on the heart precluded widespread use. It was isolated and characterised by Nagai in 1887.

Amphetamine is the generic name for the class of compounds and was first synthesised by Edeleanu in 1887. The mixture of stereoisomers was marketed as **Benzedrine** in 1932 as a decongestant. The pure (S)-stereoisomer was first prepared by Alles in 1929 and marketed as **Dexedrine** in 1937. **Methedrine,** as the most active (S)-stereoisomer was prepared by Emde in 1929. It has enhanced uptake across the blood-brain barrier.

MDA (3,4-methylendioxyamphetamine), **MDMA** or **ectasy** (3,4-methylendioxy-methamphetamine) and **DOM** (2,5-dimethoxy-4-methyl-amphetamine) are designer amphetamines used widely in the club scene.

methylphenidate (Ritalin)

Methylphenidate or **Ritalin** is widely used as a drug to treat attention-deficit hyperactivity disorder (ADHD).

All of the amphetamines enhance the release of the neurotransmitters noradrenaline, dopamine and to a lesser extent serotonin (5-HT) in the CNS. They are stimulants of locomotion, and cause excitement and euphoria and may lead to anorexia. Their hallucinogenic effects can be correlated with their affinities for 5-HT$_2$ receptors.

FLY AGARIC

ibotenic acid   **muscimol**   **muscarine**   **GABA**

Ibotenic acid and **muscimol** are the two main psychotropic constituents of fly agaric, with the former being converted into the latter (by decarboxylation) during metabolism. **Muscarine** is an agonist at one type of acetylcholine receptors – the so-called muscarinic receptors. Nicotine acts as an agonist at the other archetypal class termed nicotinic receptors.

GABA (gamma-aminobutanoic acid) is the main inhibitory neurotransmitter in the brain.

ABSINTHE

beta-thujone   **fenchone**

Beta-thujone is one of the major constituents of absinthe, though there may also be similar quantities of **fenchone** depending on the plants that have been used in the manufacture of the liquor. Thujone was always blamed for the alleged psychotropic properties of the drink, but this has now been largely disproved.

Subject Index